Explaining Reproduction

Student Exercises and Teacher Guide for

Grade Nine Academic Science

Mike Lattner *Algonquin and Lakeshore Catholic District School Board*

Jim Ross *The University of Western Ontario*

rosslattner
educational consultants *London Ontario Canada*

Lattner, Mike, 1957- ; Ross, Jim, 1952-

Library and Archives Canada

Cataloguing in Publication

Explaining reproduction : student exercises and teacher guide for grade nine academic science / Mike Lattner, Jim Ross.

1. Reproduction--Study and teaching (Secondary)
2. Human reproduction--Study and teaching (Secondary)
I. Ross, Jim (James William), 1952-   II. Title.

QH471.L38 2004   571.8   C2004-906481-9

Authors   Mike Lattner
                        Jim Ross

Printer   CreateSpace
Cover Design   Images, London, Ontario, Canada

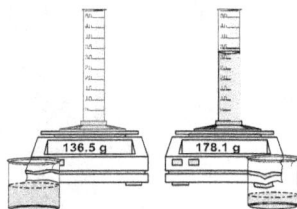

ISBN   978-1-897007-02-0

Offices   London Ontario Canada

To teachers, parents and students everywhere who desire to bring about new ways of understanding the world.

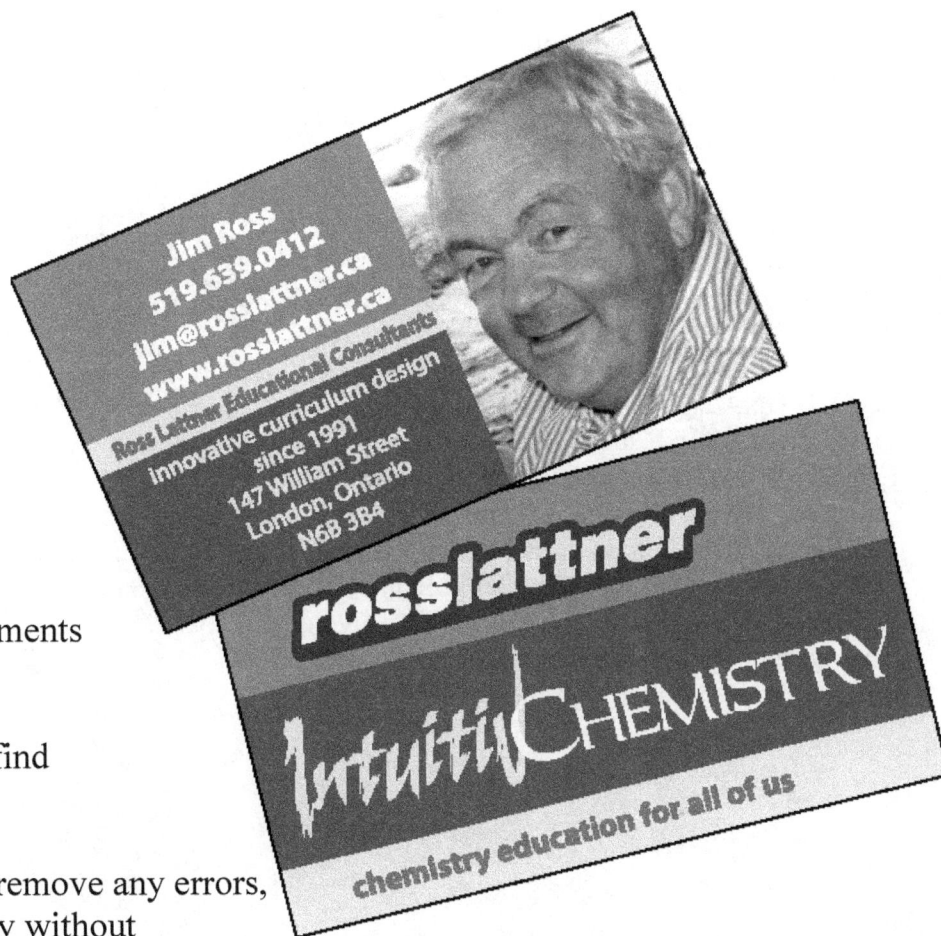

Jim Ross
519.639.0412
jlm@rosslattner.ca
www.rosslattner.ca
Ross Lattner Educational Consultants
innovative curriculum design
since 1991
147 William Street
London, Ontario
N6B 3B4

rosslattner
Intuitiv CHEMISTRY
chemistry education for all of us

We welcome your comments and suggestions.

Let us know what you find most useful.

We've worked hard to remove any errors, but don't let a day go by without letting us know if you find one.

Stay in touch.

Jim Ross

Our thanks to all of the wonderful people at the Faculty of Education, the Unversity of Western Ontario.

Special thanks to Jon McGoey, a great biology teacher and a great friend

Explaining Reproduction

Table of Contents

1: Teaching Reproduction 1

Unit Planning Notes: 2
Lab 1.1: Using the Microscope 4
Lab 1.2: Drawing from the Microscope 6
Lab 1.3: The Structure of a Plant Cell 8
Lab 1.4: The Structure of an Animal Cell 8
Quiz 1.5: Cell Structure and Function 10
Activity 2.1: Cell Growth and Reproduction 12
Activity 2.2: Nuclear Reproduction (Mitosis) 12
Activity 3.1: DNA, Genes, and Chromosomes 14
Activity 3.2: DNA During Mitosis 16
Lab 3.3: Observing Mitosis in Prepared Slides 18
Lab 3.4: DNA and How It Works 20
Lab 3.5: Genes and How They Work 22
Lab 4.1: Asexual Reproduction in Plants and Animals 24
Lab 4.2: Viewing Asexual Reproduction in Prepared Slides 26
Lab 4.3: Sexual Reproduction in Animals 28
Activity 4.4: Sexual Reproduction in Plants 30
Activity 4.5: The Human Female Fertility Cycle 32
Activity 4.6: Human Reproduction and Development 32
Project 4.7: Reproductive Biology 34

Explaining Reproduction

Table of Contents

2: Explaining Reproduction 36

Introduction: Three Theories to Understand Reproduction 37
Lab 1.1: Using the Microscope 39
Lab 1.2: Drawing from the Microscope 41
Lab 1.3: The Structure of a Plant Cell 43
Lab 1.4: The Structure of an Animal Cell 45
Quiz 1.5: Cell Structure and Function 47
Activity 2.1: Cell Growth and Reproduction. 51
Activity 2.2: Nuclear Reproduction (Mitosis) and Cell Division 54
Activity 3.1: DNA, Genes and Chromosomes 57
Activity 3.2: DNA During Mitosis 59
Lab 3.3: Observing Mitosis in Prepared Slides 63
Lab 3.4: DNA and How It Works 65
Lab 3.5: Genes and How They Work 67
Nucleotides to Build DNA models 69
Eighty Amino Acids 70
Lab 4.1: Asexual Reproduction in Plants and Animals 71
Lab 4.2: Viewing Asexual Reproduction in Prepared Slides 73
Lab 4.3: Sexual Reproduction in Animals 75
Act 4.4: Sexual Reproduction in Plants 77
Act. 4.5: The Human Female Fertility Cycle 79
Lab 4.6: Human Reproduction and Development 81
The Genetic Code: The DNA Codons for Amino Acids. 83
Project 4.7: Reproductive Biology 84

Appendix: Laboratory Safety 87

1: Teaching Reproduction

Title:	Explaining Reproduction
Time Allocation:	27.5 hours (22 periods of 75 minutes each)
Authors:	Mike Lattner and Jim Ross
Date:	November 2004

Unit Description: Multicellular organisms grow, repair tissue, and pass on hereditary material by forms of cell division and differentiation. These processes are explored in this unit of study with emphasis on the genetic information and its faithful and complete transmission between generations. We explore the structure and role of DNA, its manipulation by scientists, and the societal effects of this technology.

The unit itself is subdivided into four major sections. Each section will take a little more than one week to complete.

1. Structure and function of the plant and animal cell. Basic use of the microscope to observe the plant and animal cells. This might be review for many students.
2. The observable features of cellular reproduction, which are divided into two distinct phases. The first phase is cell reproduction, which involves the whole life cycle of one cell. The second observable phase is nuclear reproduction, which is widely known as mitosis.
3. Reproduction at the molecular level. What's going on down there during mitosis and meiosis? What can go wrong?
4. Laboratory activities in which students examine some forms of reproduction.

At the end of each section is a thorough quiz

Strand: Biology

Expectations: Overall Expectations: BYV.01-.03
Specific Expectations: BY1.01-.10; BY2.01 - .10; BY3.01 - .04

Adolescents have a natural intelligence that allows them to understand satisfactorily a number of issues surrounding heredity. The structure of their thought, however, is probably not scientific.

One goal of teaching for understanding is to use the students' innate desire to "make sense of" problems in their world. At the same time, we want to challenge their reasoning, and move students toward the reasoning used in the scientific community.

Unit Planning Notes:

This unit extends the chromosome theory of genetics. Chromosomes are discrete structures which occur in eukaryotic cell nuclei and contain one or two DNA double helices (in their replicated and unreplicated forms, respectively), and are associated with protein, especially when condensed. During cell division, chromosomes become readily visible. Their movement during mitosis, and to a lesser degree, during meiosis, will be studied. Students will produce representations of chromosomes undergoing mitosis. These representations have been carefully arranged to correspond to the ways that young people appear to think. Asexual reproduction in plants and simple animals should be seen as an extension of cell division. Sexual reproduction, involving formation and exchange of gametes, will be investigated in plants and animals, with special emphasis on the role of hormones in human reproduction. Subsequent topics include the role of DNA as a pattern for cellular activities. Since the function of DNA is critically linked to its structure, students will build models of DNA and use this to produce a protein. Hence, the concept of the importance of the genetic material as the primary pattern of cellular activities is promoted in this unit.

Prior Knowledge Required

This unit assumes that the student is familiar with some of the more common structures and organelles found in plant and animal cells, to the extent that they can recall the structure and basic function of the organelles. The student is also expected to recognize that cells in multicellular organisms need to reproduce more cells for growth and to repair tissues. A knowledge of DNA as the genetic material is not expected. Because of its emphasis upon explanation, this unit expects that the student is capable of writing coherent sentences and paragraphs.

Teaching and Learning Strategies

The focus of science is not nature itself. The focus of science is our *shared representations of nature*. Accordingly, two learning strategies are emphasized. Students are expected to explain (share) their concepts, in both *pictorial representations* and in *sentences*. In addition, students are expected to gradually master a small set of *theoretical propositions*, and then to increasingly represent their arguments using those theoretical propositions.

Assessment and Evaluation

A variety of strategies and instruments will be used throughout this document.

Introduction

Students don't "learn" what comes into them via their senses.

Students "learn" what comes out of themselves via their representations.

The student learning goals of this unit are: To describe, analyze and apply cell theory to cell division and reproduction; To evaluate recent developments in reproductive technology with implications for social decision making.

The theories that we use consist of five or six simple statements which can be used to explain things that happen around you. They even help you to predict things you have never seen!

1. **The Cell Theory** What is the underlying structure of living things? It is widely accepted among scientists that all living things are made of cells. For example, you are made of approximately sixty trillion cells. Your earlobe might have five hundred million cells in it, with perhaps hundreds of distinctly different kinds of cells! The cell theory involves these four key ideas:
   1. **All living things are composed of cells**
   2. **Cells are the basic structural unit of life**
   3. **Cells are the basic functional unit of life**
   4. **All cells come from pre-existing cells**
   The emphasis is clearly on eucaryotic cells.

2. **Cell Division and Reproduction** Eucaryotic cells have similar structures, with several common features. We will consider the roles of some cellular structures in cell division and consider their influence on reproduction. Six key ideas about cell division are:
   1. **Multicellular organisms are able to grow and repair tissue by a process called cytokinesis or cell division.**
   2. **Cell division is preceded by nuclear division (*mitosis*).**
   3. **Mitosis leads to the formation of daughter cells that are genetically identical to the parent cell.**
   4. **Mitosis occurs in almost all somatic (body) cells.**
   5. **Meiosis is the process of cell division that results in production of gametes (e.g., sperm and eggs).**
   6. **Meiosis occurs only in the sex organs.**
   Note that these ideas deal with how cells create intact copies of DNA, but do not deal with the function of DNA itself.

3. **DNA and Reproductive Biology** DNA is the genetic material contained in the nucleus of eucaryotic cells that is the pattern for all cellular processes. *Asexual reproduction* leads to daughter cells whose DNA is identical to the parent's DNA. *Sexual reproduction* produces offspring with DNA that is different from the parent's. Humans have come to understand that by manipulating the genetic material, by engineering new DNA, we can introduce new genes into the genetic material and actually produce new life forms. These advances in reproductive biology can result in tremendous benefits. Students must not be blind to the threats we can see in these technological developments.

Lab 1.1: Using the Microscope

Microscopes make it possible to see that living things are made mostly from cells.

Learning Expectation 1.02: Use a microscope at an appropriate level of magnification to locate and view objects on a slide.

Proper care and use of the microscope is an essential science skill.

Students don't just *see* what's under the microscope. They *interpret*. A novice student's microscope drawings change over a period of time. The case can be made that the student is not simply becoming a more skillful artist, but a more conceptually adept interpreter. If that is the case, teachers must provide both time and adequate opportunities to discuss the meaning of students' drawings.

Pedagogical Issues
Viewing objects under a microscope for perhaps the first time is extremely motivational for most students. However, proper use and care of the microscope is critical to any study of cells and cell division. It is important to make students realize that this valuable tool can easily be damaged by misuse. Encourage individual accountability by assigning students to specific microscopes using a numbering system. Enlist a volunteer to collate list of students and the microscopes you have assigned. Make it clear that focusing a microscope is an essential skill in science that all students must master. To enhance your students appreciation for the wonder of what can be seen using microscopes, use a video camera that records images through the lens of the microscope (e.g., *Videoflex* ®, or a simple home made device constructed from an ordinary camcorder and tubular plumbing insulation). These video units make it possible for you to show on a television monitor anything that you can bring into focus using a microscope.

Science Issues
Throughout the unit, student understanding of cell theory and cell division will be influenced by how they interpret what they see under the microscope. It will not be possible to see all organelles with student grade light microscopes. However, most student grade microscopes can be used to bring prepared slides with stained chromosomes into sharp focus. It is essential that students realize that the limits of the light microscope's ability to focus and resolve small objects places a limit on them as observers to understand what is being studied.

Cell Structure and Function

You will need to monitor your students very closely to ensure that each one can prepare a wet mount and bring it into focus with the microscope.

This exercise provides an opportunity for students with more experience to coach those who have limited microscope skills.

The Learning Activity

In this lab, students will learn the parts of a microscope and prepare a wet mount which they will use to practice focusing the microscope. This will be the first time many students use a microscope. It is important that the exercise is enjoyable and motivational in addition to being educational. Students will use their text book to label a diagram of a microscope to show the parts and to explain the function of these parts. Next, students are expected to study these diagrams so that they can repeat the exercise (possibly as homework) from memory. Finally, students will prepare a wet mount and learn to focus the object under the microscope.

A convenient specimen for use in this lab is part of a picture cut from a newspaper. Students will witness that the image is composed of a number of dots that we resolve to be an image.

Equipment, Preparation and Resources

12 microscopes (students working in pairs)
12 slides and cover slips
newspaper with photographs

It is essential that microscopes are in good working order prior to the lab. Check them in advance.

Categories:
Knowledge:
Inquiry:
Communication:
Applications, Extensions:

Assessment and Evaluation
check answers to students' labeled drawing of microscope parts
ability to make a wet mount and bring it into focus
clarity of written work

Lab 1.2: Drawing from the Microscope

The focus of science is not the study nature itself.

The focus of science is the study of our *shared representations of nature.*

A useful microscope drawing technique: place a blank sheet of 8.5 × 11 paper to the right of the base of the microscope. While keeping both eyes open, look through the 'scope with the left eye. Draw both the circle of view, and the specimen itself.

Demand that the students include a written description of what they have drawn. Kids with poor artistic ability can provide an adequate written representation of what they believe they saw.

Pedagogical Issues
Drawing from the microscope is an important way for students to demonstrate learning. For example, an obvious difference between plant and animal cells is the presence of a cell wall in plants. By representing a cell wall in a drawing of a plant and not an animal cell, a student is able to reinforce this conceptual difference. In addition, it is important for students to realize that the actual size of objects viewed through a microscope does not change; it is the size of the image, and the field of view that changes at increasing levels of magnification. The use of millimetre graph paper works well as a method to illustrate the size of the field of view.

Science Issues
Students' representations of specimens viewed in the microscope demonstrate their understanding of the underlying structure of the specimen. As such, we should recognize that microscope use encourages deeper understanding of nature.

Cell Structure and Function

Science and Pedagogy

Field of view on a typical student microscope:

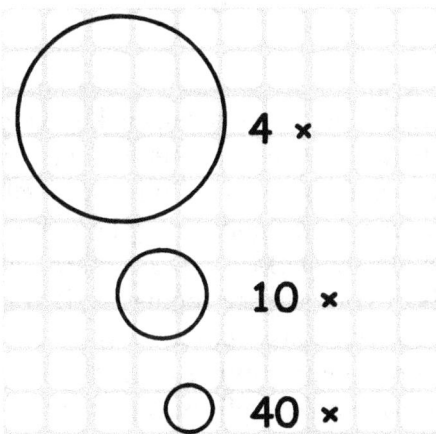

4 ×

10 ×

40 ×

Field of view 40× (4 mm)

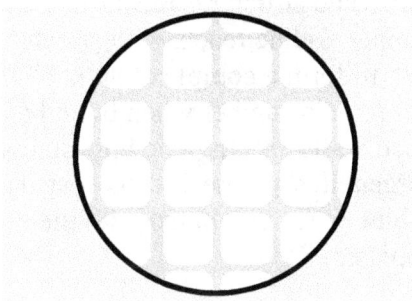

Field of view 400× (0.7 mm)

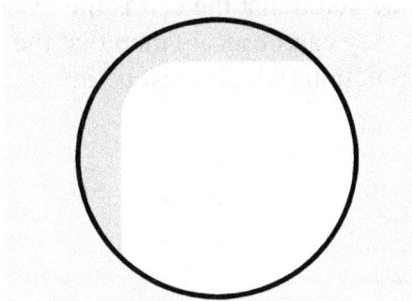

The Learning Activity
In this lab, students prepare a wet mount of a small piece of millimetre graph paper. Their drawings of the wet mount at low, medium, and high power should be completed in the space provided in the lab exercises page. Students will use the number of lines that appear in the diagram as their measurement of the field of view of their microscope.

Good copies of drawings should be completed on separate, full sheets of blank paper. At the end of the lab, students should have three drawings of the graph paper.

Students should complete all questions in the **Lab manual**.

Equipment, Preparation and Resources
12 microscopes (students working in pairs)
12 slides and cover slips
medicine droppers (to prepare the wet mount)
blank paper

Categories:
Knowledge:
Inquiry:
Communication:
Applications, Extensions:

Assessment and Evaluation
clarity and level of detail demonstrated in drawings
careful preparation of wet mount and use of microscope
quality of presentation

© Ross Lattner Publishing   7   www.rosslattner.ca

Lab 1.3: The Structure of a Plant Cell
Lab 1.4: The Structure of an Animal Cell

This is essentially a review exercise. The topic of cell structures is part of the Grade 8 curriculum. However, we include it here because students may not have used microscopes to view plant and animal cells in elementary school.

The individual cell can be considered as a system. The cell can also be considered to be one component of larger systems.

The cell membrane serves as a boundary between the cell and its environment.

Within the cell membrane are found the equipment to make proteins, stockpiles of fuel, and the proteins made by that cell for its own purposes.

Pedagogical Issues

A key difficulty students find in drawing cells is knowing what are the limits of one cell. Many of your students will not be able to delimit a single cell in the onion epidermal tissue. Focus on this concept by demonstrating how the cell walls of onion cells meet in a block like structure somewhat like the walls of your science room (i.e., if your science room is concrete block construction). Just as the walls of your science room contain your lab and its equipment, the walls of many onion cells make up the epidermis and help hold together the onion.. The use of stain (i.e., iodine for onion cells and methylene blue for cheek cells) brings out some of the internal structures of cells. Different organelles and structures react differently with the stain. For example, iodine reacts with starch, found in some vacuoles, to form a complex that is blue-black in appearance, but remains amber when bound to the cellulose found in cell walls. Methylene blue stains the nucleus and other organelles of cheek cells very dark blue, but remains light blue when combined with the components of the cytoplasm.

Science Issues

Research in science education indicates that it may be easier for students to understand that the cell is the basic unit of structure (which they can observe) than that the cell is the basic unit of function (which must be inferred from experiments).

Stress the presence of a darkly stained nucleus and the irregular shape of the cheek cell.

The Learning Activity

This activity can be done immediately after the quiz or the following day. Depending on your class, students should be able to complete both drawings in one 75 min period.

Follow up the drawings with the exercises identified in the Lab Manual. Students should use their text books as a reference to help them complete these exercises.

The presence of the cell wall gives onion epidermal cells a more regular appearance.

Equipment, Preparation and Resources

12 microscopes (students working in pairs)
12 slides and cover slips
fresh onion
scalpels
iodine staining solution (prepared in advance)
methylene blue staining solution (prepared in advance)
toothpicks
blank paper

Categories:	Assessment and Evaluation
Knowledge:	quality of answers to questions
Inquiry:	ability to prepare and focus slides
Communication:	clarity and level of detail in diagrams
Applications, Extensions:	

Quiz 1.5: Cell Structure and Function

Pedagogical Issues

Although the main student learning goal of this unit is knowledge of the processes involved when multicellular organisms reproduce, this complex set of objectives cannot be met without a basic understanding of cell structures. This set of quiz items provides a simple set of assessment instruments to help determine whether students have the basic background knowledge.

You may use these as a pre-test, to determine the range of student achievement from Grade 8, before deciding whether to use the activities in this section. It is possible that only some of these activities would be useful to your class.

Alternatively, you may use it as a post-test to check student learning in this section.

Science Issues

Items 1 - 8 deal particularly with use of the microscope, and making representations of objects seen under the microscope. Items 9 - 12 deal with the basic structure and organelles of plant and animal cells.
Items 13 - 16 question student integration of the Cell Theory with the Particle Theory, and in particular with the very large particle that we call DNA. These may provoke some discussion, and they may provide a little help in determining what interventions you decide upon in the next section.

The Learning Activity

These quizzes can be used profitably in several ways:

Daily Pop Quiz. Did the kids do the homework? Did they understand it? You can pop one of these questions on the class the day after the lesson, and quickly assess problems.

Daily Practice Quiz If half the class could do it on Tuesday, can they improve by Thursday?

Discussion Generator Some questions and responses can generate controversy in the classroom. When students are required to explain their beliefs, some very fruitful learning situations can develop

Question on a later summative test Feel free to use any of these quiz items on a summative test. Students respond more confidently to structures they have seen before.

You may well find this kind of item restrictive. By all means modify questions to suit your own classroom.

Equipment, Preparation and Resources

The Grade Nine Daily Quizzes in their lab manuals.

Categories:
Knowledge:
Inquiry:
Communication:
Applications, Extensions:

Assessment and Evaluation

These are basically knowledge items for interim assessment.

Activity 2.1: Cell Growth and Reproduction
Activity 2.2: Nuclear Reproduction (Mitosis)

Which is more supportive of meaningful learning at the introductory level:

mastery of the definitions
or
"getting the general idea?"

Pedagogical Issues   It has been said that the number of new science words introduced to a student in a high school science course exceeds the expected vocabulary of a comparable French course. Students in a science course, however, are expected to master not only the new words, but entirely new concepts as well. This load is extremely demanding, confusing and discouraging for a typical Grade Nine or Ten student.

The purpose of this set of activities is to provide some simple data that students can organize into a coherent representation of these complex phenomena, without having to master a set of abstract definitions.

The process of cell growth and replication is smooth and continuous. There are no naturally occurring stages or steps. The definitions have been introduced by scientists to "tell the story" of cellular reproduction. The words *interphase*, *prophase*, *metaphase*, *anaphase*, *telophase* and *cytokinesis* need not have any more abstract associations than those required to meet the student need to "tell the story."

Where do definitions come from? Do they arise in nature itself? Of course not.

Definitions are constructed by a community of scientists for their utility at marking off the boundaries of an idea.

Certainly non-experts need to respect the definitions of experts. But in education, the *value* of a definition is its usefulness in achieving an educational goal.

Science Issues   Scientific definitions are constructed by groups of scientists to meet certain needs. When you look through the microscope, you see pretty much what the first scientists in this field saw. They, however, did not have access to the recent enormous advances in molecular biology.

The activities and labs following this activity attempt to clarify some of these processes from the point of view of molecular biology, using a minimum of technical terms.

The Learning Activity

Cell Growth and Reproduction
In pairs, or small groups, students cut out the cards with the pictures of cell growth and division. They then arrange the cards in an order that makes sense to them. They must be prepared to explain their thinking the story of cell growth.

Nuclear Reproduction (Mitosis)
This activity has the same basic structure. In this case, the students must contend with the "fuzzy boundaries" between the traditional stages of mitosis. At the end of the exercise, they must "tell the story" of mitosis in simple terms.

Equipment, Preparation and Resources
Cards and instructions in the lab manual
Scissors
Glue
Markers or coloured pencils

Categories: **Assessment and Evaluation**
Knowledge: Are the cards in a coherent order?
Inquiry:
Communication: Quality, clarity of "the story of mitosis"
Applications, Extensions:

Activity 3.1: DNA, Genes, and Chromosomes

Pedagogical Issues On the surface, this objective appears quite simple. However, in order to become able to "talk the talk," or "think the thought," a student is required to:

> memorize more new vocabulary in one day than they would be required to do in a French class,
> *and* to learn completely novel concepts, which presumably they would not have to do in a French class
> *and* to relate these novel concepts to their existing knowledge
> *and* to avoid the misconceptions which naturally arise when a novel concepts is being worked out by the patterns and processes of everyday thinking.

allele
centromere
chromatid
chromatin
condensed chromosome
diploid
DNA
gene
genetic
haploid
homologous
replicated
sister chromatids
spindle fibres
unreplicated

This simple objective is a tall order indeed.

Once again, the approach taken here is to provide the students with some material to work with, and have them make use of everyday thinking to find patterns in the material, in this case, gene alleles located upon pairs of homologous chromosomes. This task can be done without a knowledge of the words or definitions. The pedagogical objective is that the categories arising from the exercise become the basis for a working definition of the terms used.

Science Issues At the dawn of the twenty first century the entire structure of biology, from the parts of a cell up to entire ecosystems, is founded upon molecular genetics. The idea that DNA is a very large molecule with useful patterns in it, capable of self-replication, is an enormously simplifying notion. Simplifying, that is, in the sense that the idea provides one comprehensive structure that ties everything together.

Is it possible to get students close to that contemporary understanding without tripping over the complicated jargon?

Genes, Chromosomes and DNA

The Learning Activity

In this exercise, the students:

1. **Study** the strands of DNA in the lab manual. Which pairs of strands appear to be related?

2. **Identify** the "junk DNA," and mark it all with one colour of high-lighter.

3. **Find** as many genes as they can, and mark them with other colours. This resembles a word search.

4. **Study** the text book to find all of the words at left in the index of the text, or in another resource.

5. **Complete** the boxes on the back of the lab page.

Equipment, Preparation and Resources

Coloured markers or high-lighters.

Categories:	Assessment and Evaluation
Knowledge:	Correct identification of the features required
Inquiry:	
Communication:	clarity of their explanations, especially box 1 and 4
Applications, Extensions:	

Imagine being the first scientist to see those dark bodies inside a cell, through a home-made instrument

Imagine the struggle to first see them clearly through technological improvements in stains and lenses; then to recognize their shape and extension and represent them as discrete bodies; to arrange them in some kind of order; and finally to generate a set of shorthand words to enable you to communicate with your colleagues. All that without benefit of knowledge of DNA.

Our traditional pedagogical approach to introducing this topic has been to focus almost exclusively upon memorization of the shorthand.

For reasons related both to the student's own psychology and to the discipline of science, students need some explanatory system.

This exercise is an attempt to meet both of these learning needs.

Activity 3.2: DNA During Mitosis

Pedagogical Issues Our traditional pedagogical approach to introducing this topic has been to focus almost exclusively upon memorization of the shorthand developed by scientists to communicate with each other. Millions of earnest, motivated students have struggled to put *prophase*, *metaphase*, *anaphase* and *telophase* into the correct order for decades. A few have succeeded.

The main objective: *At the end of this lesson, students should be able to model the process of mitosis.* The learning of the shorthand notation, while an important objective, is subordinate to the main objective.

Science Issues

From the molecular point of view, what are those bodies that move around during mitosis? They are DNA, all coiled up. In fact, they are triply coiled:

1. The DNA is twisted into a double helix
2. The double helix is coiled into long "rope"
3. The "rope" is coiled into a supercoil

The supercoils are the structures you see under the microscope, the things that are usually called *chromosomes*. The chromosomes don't "disappear" during telophase; they uncoil, becoming so thin as to be invisible. The act of copying or of "reading" the DNA can only occur when the DNA is uncoiled.

The coiled up DNA is the structure that students see under their microscopes. When it is coiled up, it is neither being copied or being "read."

Genes, Chromosomes and DNA

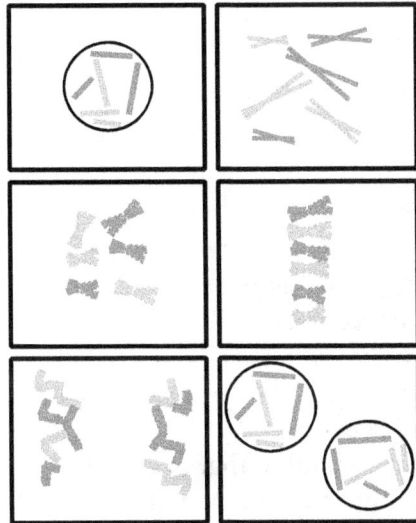

The Learning Activity In groups of 3 or 4:

Don't glue anything down, until you are sure of the step-by-step process.

1. Each group will make six models of a cell, one at each stage of mitosis and cell division, and fasten them all to a single sheet of display board to make a poster.
2. Six strands of DNA can be found on on the lab sheet. The students cut out each strip
3. Students follow the detailed instructions in the lab sheet
4. As they proceed through each step of the cell cycle, they make a new model for each step.
5. At each stage, they answer the questions in the space provided and make a diagram of the cell.
6. They should repeat the cycle several times in order to beome familiar with each part of the process.
7. Finally, they glue up the individual steps, and assemble the poster.

Equipment, Preparation and Resources

Make some extra photocopies of the DNA strands.

Scissors
glue, clear tape
6 - 8 display boards
40 - 50 sheets of paper
paper clips
thread
markers.

Categories:	Assessment and Evaluation
Knowledge:	correctly models the process of mitosis
Inquiry:	
Communication:	quality of diagrams and sentences
Applications, Extensions:	

Lab 3.3: Observing Mitosis in Prepared Slides

It is highly unlikely that students could look through the microscope and "discover" the "four stages of mitosis"

Equally problematic is the notion that students could, all by themselves, "construct" a meaningful account of the disorganized glimpses they observe under the microscope.

It seems more reasonable to believe that students are being gradually introduced into a way of publicly representing some processes which they would not encounter in everyday life.

This lab presupposes both student commitment to the hard work of constructing meaning, and teacher commitment to the delicate task of promoting meaningful dialogue.

Pedagogical Issues   Having constructed a representation of the basic processes of mitosis, students will be more able to interpret what they see real cells. Unlike the representation of Activity 3.2.3, however, the cells under the microscope are not neatly organized, do not appear to follow a particular order, etc., so a considerable degree of interpretation is necessary.
It is important that students remember that mitosis is essentially a mechanism to faithfully copy and distribute chromosomes so that two daughter cells, which are genetically identical to the parent cell, result from the process.

Science Issues   During the daily activities of organisms, cells are constantly being replaced as necessary. A single fertilized human zygote undergoes repeated cell divisions to become an independent organism. How many cell divisions, each one doubling the number of daughter cells, would be necessary to produce the sixty trillion cells in the human body? How many cells must die during this process?
The length of time required to complete a cell cycle varies greatly, depending on the organism, and on the type of cell. For example, human cells grown in culture, require approximately 24 h to complete one cell cycle. Onion root cells can divide in about half that time.

Cells prepared for the purpose of viewing mitosis are often stained with acetocarmine to highlight the appearance of the chromosomes.

Genes, Chromosomes and DNA

Prepared slides will show many cells in different stages of mitosis.

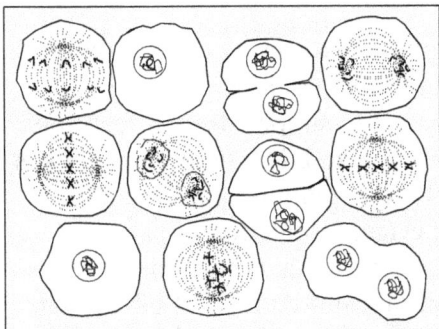

You can make your own video camera for microscope viewing. Video cameras are sufficiently sensitive to make good images. Use a piece of polyurethane plumbing insulation to connect the camera and the microscope tube. Use standard video cable to connect the camera to the monitor.

The Learning Activity Begin this lab with a review of the stages of mitosis and the key events that occur during each stage.

To assist students in identifying what to look for, use a video camera capable of viewing through a microscope (e.g., Videoflex®). Show students that it will be necessary to search around the slide to find all mitotic stages.

Students are expected to make clear and detailed drawings of prophase, metaphase, anaphase and telophase in the space provided in the Lab Manual.

Equipment, Preparation and Resources
(sufficient materials for a class of 24, working in pairs)
12 microscopes
12 prepared slides of cells undergoing mitosis (e.g., white fish embryo or *Alium* sp.)

Categories:	Assessment and Evaluation
Knowledge:	answers to activity questions
Inquiry:	
Communication:	clarity of response
Applications, Extensions:	

Lab 3.4: DNA and How It Works

> "Students do not learn what goes into them via their senses.
>
> Students learn what comes out, the things they express in their representations."

Pedagogical Issues Students can *recognize* representations they have seen before. This is in fact, a kind of learning. Simple recognition is not sufficient to support further learning. Students can quickly learn to recognize DNA in text readings or the internet, but may not be able to explain why its structure permits inheritance, the near- perfect replication of patterns.

This simple model building exercise takes students through the structure and the replication of a short strand of DNA. The very act of using one's hands to perform an operation like this means that many kinds of mental representations must occur simultaneously in the brain. The likelihood of learning may be increased.

How could you, as a classroom teacher, determine wether model building or simple reading was more effective at supporting long-lasting conceptual learning?

Science Issues At any temperature, molecules of air must be traveling at a little more than the speed of sound. There is no other way for sound to propagate than by the physical movement of air molecules in a series of collisions. In fact, the speed of air molecules at room temperature is about 450 m/s, much faster than a high-powered rifle bullet.

Water molecules are lighter than air molecules. At the same temperature (kinetic energy) water molecules are moving even faster than air molecules, perhaps 500^+ m/s.

> How quickly does a replication proceed?

Imagine DNA existing for years in each cell in your body, remaining unchanged throughout the incredible pounding by the water molecules. The nucleotide chain must be strong enough to withstand this pounding. The C-G and T-A base pairs are held together only by hydrogen bonds. How can they withstand such a constant beating? The double helix structure keeps the strands from wandering apart. When the double helix is opened, the strands are readily broken up in the thermal environment.

Genes, Chromosomes and DNA

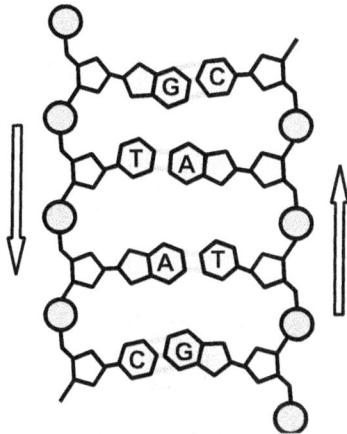

Structure of DNA double helix showing adenine bonded to thymine and guanine to cytosine. There are three hydrogen bonds between Guanine and Cytosine, and only two hydrogen bonds between Adenine and Thymine.

The Learning Activity In this activity, students cut and paste paper nucleotides together to make one half of a DNA double helix. Next, students will match their DNA sequence with a complimentary strand, thus producing double stranded DNA.

1 Make the Chain Use the accompanying paper DNA template to make a chain of 10 nucleotides. Note that the P-D-P-D chain (Phosphate-Deoxyribose-Phosphate-Deoxyribose...)

2. Make the complementary strand. Make the opposite change, with the G-C and A-T cross lings as shown in the diagram at left.

3. Unzip the complementary chain into two. DNA is able to unzip into two strands, by separating right down the middle, breaking the A-T bonds and the C-G bonds.

4. Rebuild two new strands by repeating Step 2 on both of the single strands.

5. Compare the two new double strands to see if indeed they are the same.

Equipment, Preparation and Resources
(sufficient for a class of 24)

paste
blank paper
nucleotide cut outs (see **Lab Manual**)

Be sure to make 24 copies of each of the 4 nucleotides prior to this activity.

Categories:
Knowledge:
Inquiry:
Communication:
Applications, Extensions:

Assessment and Evaluation
answers to Questions for Later..

Quality of the DNA models

Genetic traits are expressed largely as a result of enzyme activities. The hereditary patterns exist inside the organisms as patterns of DNA of their chromosomes.

An altered gene, if not repaired, can be passed on to every cell that develops from it.

Gene mutations can be caused by such things as radiation and chemicals. When they occur in sex cells, the mutations can be passed on to offspring. If the mutation occurs in a body (somatic) cell the change will be evident only in the affected generation.

Fortunately, our cells have very efficient systems for recognizing and repairing mutations. Without these mechanisms, the effects of U-V light, for example, could cause severe damage to our skin cells.

Lab 3.5: Genes and How They Work

Pedagogical Issues When we say "the organism stores information in its DNA," students are likely to take us literally.

You have heard students say things like "the atoms want to cross over the membrane." Take them seriously. The structure of their thought requires the existence of a conscious atom, with a will. Thinking that way, the student need not consider things like energy, movement, randomness, etc. In the present case, the idea that a cell just "decides to store some information" in DNA is taken literally. The student who thinks that way is not being challenged to describe the situation in an explanatory fashion.

Science Issues Information present in DNA is actually expressed by a more complex process of transcription into RNA and translation into protein. Perhaps, at this level, it is enough to focus on gene expression in so far as it relates directly to DNA, leaving out the middle step.

The genetic pattern is not *ambiguous*. For example, if a gene has the sequence of nucleotides - adenine, adenine, adenine, adenine, guanine, cytosine - a protein would be produced consisting of the amino acid, lysine, chemically bonded to the amino acid, cysteine and no other amino acids. For each sequence of three nucleotides, only one type of amino acid is added to the protein.

However, the genetic pattern is *redundant*: any one amino acid can be coded for by more than one sequence of three nucleotides (e.g., CCC, CCA, and CCG all code for the amino acid, proline).

Genes, Chromosomes and DNA

In reality, the genetic pattern is first copied onto messenger RNA (mRNA). The pattern used in this lab has been simplified to facilitate understanding.

The simulated mutations in this activity are commonly called point mutations. In reality, X-rays affect DNA by creating many free radicals which then react with the nucleotides.

UV light can cause pyrimidine dimers to form. Imagine two pyrimidines adjacent to each other. If they are struck by a UV photon, the pyrimidines can form a dimer. Such a dimer introduces a "kink" in the double helix. If this kink is not repaired, protein synthesis is blocked from that point on.

Chemical mutagens, such as nitrous acid (HNO_2), act directly on the DNA by changing the hydrogen bonding between bases, thus changing the base at one or more points in the DNA sequence.

Mutations are changes in the hereditary message of an organism. They may result from physical or chemical damage to the DNA or from spontaneous errors during replication. Usually, mutations have a deleterious effect on cells. However, if a heritable mutation results gives the individual who inherits it even a 0.01% increase in offspring, the new gene will likely be preserved in the gene pool of the population.

The Learning Activity

In this investigation, students will make a short sequence of protein based on the information carried in simulated genes. Students must randomly write down a sequence of 30 nucleotides (e.g., AATGGCCGT, etc.). Then, using this DNA, students will transcribe the codons in the nucleotide sequence into 10 amino acids to make a small protein (polypeptide) using the genetic code provided in the **Lab Manual**. Note, the genetic code is not punctuated; the sequence is read 3 nucleotides (i.e., one codon) at a time. Students should write down the list of amino acids, in order, in the space provided. Also, they should cut out amino acids from the template, also found in the **Lab Manual**, to make an actual simulated protein. The effects of three major sources of mutational damage: high-energy radiation, U-V light, and mutagenic chemicals will be investigated by how each agent, in turn, changes the initial DNA sequence and the resulting proteins. Have students compare the initial amino acid sequence with the proteins resulting from mutation.

Equipment, Preparation and Resources
(sufficient for a class of 24, working in pairs)

At least 50 copies of the amino acid template prior to class. scissors, cellulose tape, glue sticks

Categories:
Knowledge:
Inquiry:
Communication:
Applications, Extensions:

Assessment and Evaluation
clarity of answers to *Questions for Later ...*
ability to prepare proteins from a sequence of DNA

Explaining Reproduction

Lab 4.1: Asexual Reproduction in Plants and Animals

Plants and animals resulting from regeneration have the same DNA as the parent (barring mutations), since new cells result from replication of DNA in mitotic divisions.

A cutting is a part of the plant, (a stem or a leaf), that has been separated from the parent plant, and which will regenerate a complete plant.

Cut *Planaria* as shown to find which parts can be regenerated.

Planaria cut at the head

Planaria transverse cut

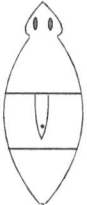

Planaria with double transverse cut

Pedagogical Issues In this lab, students use isolated parts from a plant and *Planaria* to regenerate whole individuals. The purpose of the lab is to reinforce the notion that, in general, asexual reproduction results in offspring with genetic characteristics identical to the parents. **Start this lab early in the unit to allow time for the cuttings to grow and develop.**
Stem cuttings can be made from the tips of healthy, vigorous young stems of *Tradescantia*, geraniums, or begonias. In preparing the stem cuttings, use a very sharp knife and cut below a node (where the leaf joins the branch). Include several nodes, and make the cutting approximately 5 to 10 cm long. Pinch off any flower buds and cut off the leaves to reduce water lost by transpiration. Plant the cutting to a depth of about 3 cm in moist sand. The sand must be well drained to prevent rot, and kept moist to prevent wilting.

Science Issues Regeneration in plants may occur naturally, in which case it may be considered a form of asexual reproduction. Some invertebrates - especially worms, are able to regenerate missing parts. New cells are formed as a result of differentiation and growth.

Planaria can be obtained from a biological supply relatively cheaply and maintained in the growth medium provided or in pond water. Change the water frequently to prevent contamination. No feeding is necessary.

Varieties of Reproduction

Allow several weeks for the cuttings to grow and develop.

The Learning Activity

Students will prepare cuttings of plants and *Planaria* to demonstrate regeneration and asexual reproduction.

Before the experiment, insist that the students to make a prediction of what will happen to the cut plants and animals and to write down why they think this will happen. It is important to make a commitment to some line of thought. During the days of this unit, students should be able to observe their specimens and record their observations.

At least two weeks will be required to determine if the cuttings have been successful. After the experiment, students should record their observations and attempt to explain their results.

Equipment, Preparation and Resources

(sufficient for a class of 24, working in pairs)
at least 50 *Planaria*
one or two healthy young begonias, geraniums, or *Tradescantia*
12 sharp scalpels or razor blades
sand for plant cuttings
containers for plant cuttings
petri dishes
darkened storage area for *Planaria*

Categories:

Assessment and Evaluation

Knowledge: last explanation of the Predict, Explain, Observe, Explain cycle
Inquiry: dedication to observing specimens during growth phase
Communication: clarity and quality of student explanation
Applications, Extensions:

Fission in protozoa, budding in yeast, and sporulation in fungi are examples of asexual reproduction.

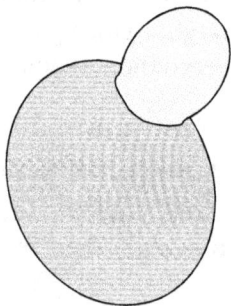

Budding in yeast. The bud is a perfect copy of the parent yeast cell, because mitosis has provided a near-perfect copy of the original DNA pattern from which the mother was expressed.

(Note the asymmetrical distribution of cytoplasm)

Lab 4.2: Viewing Asexual Reproduction in Prepared Slides

Pedagogical Issues

This lab is intended to allow students to view asexual reproduction as it actually occurs in certain organisms. Again, it is important to stress that asexual reproduction results in offspring that are genetically identical to the parent.

Science Issues

Fission involves equal division of living material and occurs in certain protists, bacteria, and algae. Budding, which occurs in yeast, involves an unequal division of cytoplasm, but an equal complement of DNA being passed to the small bud during mitosis. Spore formation, for example in the bread mold *Rhizopus*, happens under favourable conditions of heat and humidity, and results in each spore having an identical complement of DNA.

Varieties of Reproduction

Biological drawings should demonstrate at least three levels of detail. First, show the outside of cells and their environment. Second, show what is found in cells (e.g., organelles). Third, provide some structural detail of the organelles or cytoplasm.

The Learning Activity

This lab will allow students to consolidate their concept of asexual reproduction and reinforce the notion that the results of the process are offspring genetically identical to the parent.

As before, care should be taken when using prepared slides so that students do not damage the slides by raising the microscope stage so that the slide is cracked by the objective lens.

Help students to see the important details by using a microscope video camera and monitor.

Drawings should be neat and clear and show three levels of detail.

Equipment, Preparation and Resources

(sufficient for a class of 24 working in pairs)
12 compound microscopes
12 slides of **Rhizopus, Paramecium, Saccharomyces**

Categories:
Knowledge:
Inquiry:
Communication:
Applications, Extensions:

Assessment and Evaluation

clarity of responses to questions for later

detail demonstrated in drawings

Explaining Reproduction

Mitosis   Meiosis

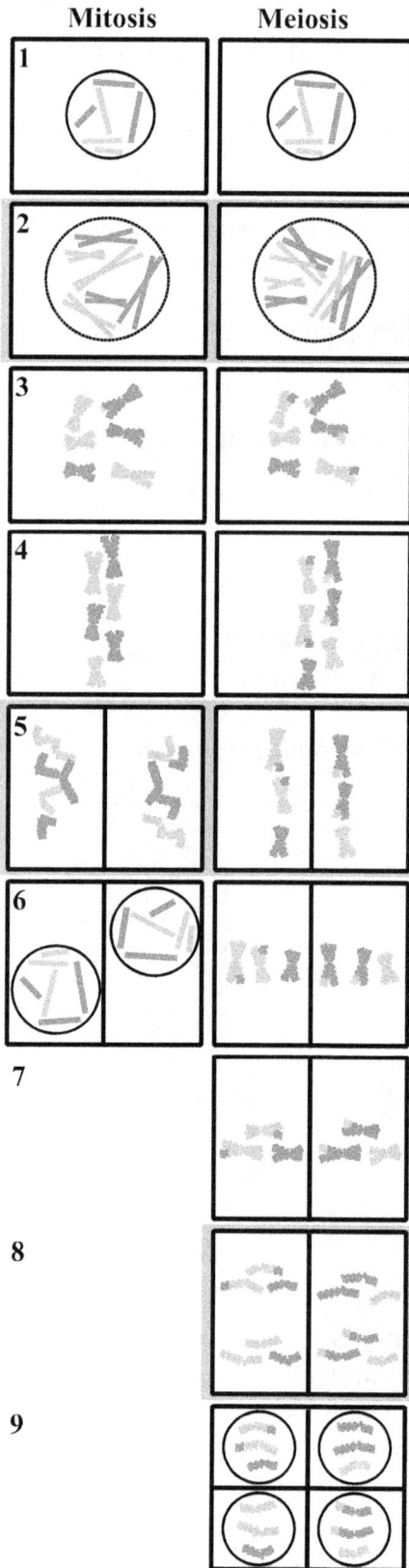

Lab 4.3: Sexual Reproduction in Animals

Pedagogical Issues Sexual reproduction is a hot topic among grade nine students. They will tend to think of it almost exclusively in terms of human sexual intercourse. It's important to emphasize that, from the genetic perspective, sex is a process which mixes DNA from two individuals.

Discussion of meiosis need not focus on the difficult terminology. The important point is that not one of the four gametes produced is identical to the others, or to the parent. Mitosis (left column in illustration) and meiosis (right column in illustration) differ the most at crossover mixing (step 2), separation of sister chromatids (step 5) and separating single chromosomes rather than pairs (step 8).

Science Issues Sexual reproduction involves the union of two cells, usually called gametes. Hence, somewhere in the life cycle of the organism, *meiosis* or *reductive division* must occur to prevent doubling of chromosomes. Meiosis involves two successive divisions, but only one stage in which the chromosome number is reduced. In most animals, gametes are monoploid (1N) as a result of meiosis.

Conjugation refers to the fusion of two cells that appear to be structurally alike, whereas fertilization is the union of two cells that are structurally different (i.e., egg and sperm).

Daphnia demonstrate forms of asexual and sexual reproduction. During summer, most are diploid females that reproduce parthenogenetically. Each female carries up to 100 eggs that develop internally and eventually develop into adult diploid females. When faced with environmental stress, male *Daphnia* are produced from some eggs. These males fertilize specialized eggs that can withstand freezing and drying out and form the beginning of next summer's *Daphnia* population.

Varieties of Reproduction

Conjugation in paramecia

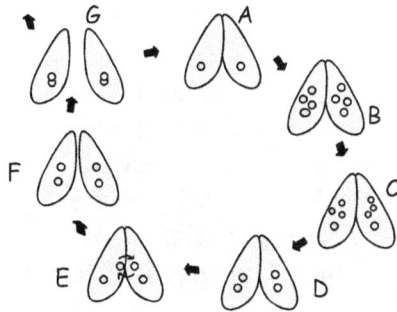

A, B the single micronucleus in each conjugant divides twice, producing four haplo-nuclei. Three disintegrate **C. D,** each micronucleus divides. **E** the cells exchange one of their micronuclei. **F, G** the micronuclei fuse resulting in two individuals with new DNA.

Daphnia can be obtained from most biological supply companies. (NOTE: you must order these live specimens in advance and specify a date for arrival at the school). They can be viewed under microscopes in depression slides to which a small amount of methyl cellulose has been added to slow them down.

The Learning Activity

Start the lab with a demonstration of meiosis using the same materials as in Activity 4.4. Upon completion of the demonstration have students complete the chart in the Lab Manual. Next, have students label the diagram of conjugation in paramecia using the description at left. If you decide to view **Daphnia** the eggs in the brood pouch of parthenogenetic females are easily visible. Have students sketch the appearance of the adult female with eggs. Answers for the remainder of the lab can be prepared in the Lab Manual or separately with the aid of a text book. Encourage students to focus on the number of chromosomes present in somatic (body) cells and in gametes. Also, note how internal fertilization is more efficient in terms of the return on the reproductive investment of a female.

Equipment, Preparation and Resources

(sufficient for a class of 24, working in pairs)
12 red pens and 12 blue pens
12 red pencil crayons and 12 blue pencil crayons
12 red 15 cm rulers and 12 blue 15 cm rulers
string and yarn (see Activity 4.4)

12 microscopes and depression slides (for **Daphnia**)

Categories:
Knowledge:
Inquiry:
Communication:
Applications, Extensions:

Assessment and Evaluation

correctly answering questions
ability to focus and view **Daphnia** (optional)
clarity of responses in **Questions for Later**

Sexual reproduction can be investigated in the filamentous alga, *Spirogyra*, and in mosses, ferns, gymnosperms, and angiosperms.

Life Cycle of a Moss

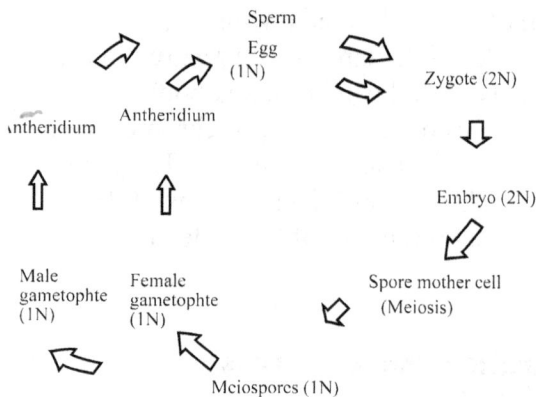

Sperm
Egg (1N)
Zygote (2N)
Antheridium
Antheridium
Embryo (2N)
Male gametophte (1N)
Female gametophte (1N)
Spore mother cell (Meiosis)
Mciospores (1N)

The ginkgo tree (*Ginkgo biloba*) is a living fossil. It is the only living form of widespread ancestral types known only in fossils, some as old as 200 000 000 years.

Tulips, daffodils, and gladioli are excellent forms to show as almost diagrammatic flowers because their pistils and stamens are very clear.

Activity 4.4: Sexual Reproduction in Plants

Science Issues *Spirogyra* can be viewed in prepared slides to show conjugation in primitive plants. This alga exists as monoploid filaments which fuse to allow exchange of monoploid isogametes and production of diploid zygospores. Draw students attention to the conjugation bridge and the genetic material that is exchanged between "active" male and "passive" female cells.

For this activity, it may be helpful to have students bring in cones from conifers (gymnosperms) and flowers from angiosperms. Students will learn that cones differ greatly in size and appearance and that not all flowers have the same arrangement of petals, stamens, and anthers. The life cycles of mosses and ferns have alternating generations of haploid (or monoploid) gametophyte and diploid sporophyte stages. Contrast these with the life cycles of gymnosperms which feature a much reduced gametophyte stage that is contained in the male and female cones of usually the same tree. In flowering plants (angiosperms) the conspicuous stage is the diploid sporophyte, while the monoploid gametophyte is reduced within a pollen grain and ovule.

If Ginkgo trees are known to be present in your area, they could form the basis of an interesting study. Ginkgo are dioecious; pollen trees have catkins containing microspores, and paired ovules are found on short stems of seed trees. Microspores are carried by the wind to ovules, where they develop into gametophytes. Sperm cells are flagellated and must swim through pollen tubes to reach egg nuclei. Ginkgo, along with cycads, are the only seed plants with flagellated male gametes.

This activity could be enhanced with the presence of fresh flowers. Commercial flower shops have interesting examples at virtually all seasons.

The Learning Activity

This lab involves the use of microscopes or micro viewers to observe prepared slides of conjugation in filamentous alga. Caution students on careful use of the microscope to observe transfer of the genetic material.

Next, present the life cycle of mosses that appears on the previous page and in the **Lab Manual**. Highlight the alternation of the gametophyte and sporophyte generations. Also, contrast the life cycle of the moss with that of the fern. The gametophyte of the moss is the conspicuous stage, familiar as the green plant; the sporophyte grows on top of the gametophyte and is parasitic on the gametophyte. In ferns, the sporophyte plant is the conspicuous frond. Under fronds are found sori that contain spores - the first stage of the gametophyte generation.

Students can use their text book to produce life cycles of angiosperms and gymnosperms.

Equipment, Preparation and Resources
(sufficient for a class of 24, working in pairs)

12 microscopes
12 prepared slides of **Spirogyra**
posters of the life cycles of mosses, ferns, gymnosperms, and angiosperms

Categories:
Knowledge:
Inquiry:
Communication:
Applications, Extensions:

Assessment and Evaluation
correct life cycles of plants

clarity and detail in drawings

Meiosis in the germinal cells of testes and ovaries produces gametes with different haploid sets of chromosomes. Random union in fertilization produces offspring with chromosome combinations different from either parent.

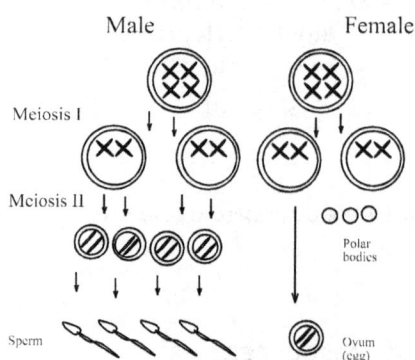

Male   Female

Meiosis I

Meiosis II

Polar bodies

Sperm   Ovum (egg)

Reproductive hormones also affect males. FSH stimulates spermatogenesis. LH stimulates secretion of testosterone, which stimulates development and maintenance of male secondary sexual characteristics.

Activity 4.5: The Human Female Fertility Cycle
Activity 4.6: Human Reproduction and Development

Science Issues

Focus on the random assortment of chromosomes that results in fertilization. This leads to variation. Fertilization happens after sperm is transferred to the vicinity of a viable ovum during copulation. Egg and sperm cells have finite and short life spans, so timing is critical. In humans, gametes must unite within hours of release for conception to occur.

Control of the hormones involved in human reproduction begins at the hypothalamus which produces releasing factors that stimulate production of the gonadotrophic hormones, Follicle Stimulating Hormone (FSH) and Luteinizing Hormone (LH), by the pituitary gland. These hormones stimulate the growth of the egg in a follicle in the ovary, ovulation, and the production of the corpus luteum. Ovulation occurs when LH and FSH levels peak. Estrogen and progesterone are produced in the ovary by the follicle and corpus luteum. Together, these hormones prepare the lining of the uterus to accept a fertilized egg. When estrogen and progesterone levels fall, a feedback loop affects the production of further releasing factors by the hypothalamus. The pituitary can again initiate production of FSH which triggers menstruation.

Sperm are most efficiently produced at a temperature that is slightly below core body temperature. In humans, and in other mammals, the scrotum holds testes outside the main cavity of the body, reducing their temperature.

At birth, the ovaries in a baby girl already contain thousands of partially developed egg cells called primary oocytes. Each is surrounded by a layer called a follicle. About once a month, FSH triggers maturation of usually a single egg. Multiple fraternal births result either from more than one egg maturing in a single ovary, or from stimulation of follicles in both ovaries.

The Learning Activity

This lab consists of two activities. In the first, students will relate meiosis to reproduction. Focus on the relatively large number of male gametes produced in comparison to the one (usually) egg produced in each menstrual cycle.

In the second part, students will graph changes in levels of hormones or draw changes in the ovary and uterine lining during one complete ovarian cycle. In this activity, focus on the relative levels of hormones and the changes in the follicle, ovum, corpus luteum, and uterine lining that coincide with changing hormonal levels. Instruct students to use different colours to show the changes.

Answers for the questions should be prepared neatly on a separate sheet, although there is space in the **Lab Manual** for short notes.

Equipment, Preparation and Resources

Students will need to use their text books for relative levels of the reproductive hormones during a menstrual cycle. If available, have on hand a large wall chart showing the interactions of the reproductive hormones. During the next class, students will focus on conception, pregnancy, development, and birth.

Categories:
Knowledge:
Inquiry:
Communication:
Applications, Extensions:

Assessment and Evaluation

correct responses to questions

clarity and level of detail demonstrated in graphs

Project 4.7: Reproductive Biology

Pedagogical Issues

Writing a short essay is always a challenge to students. A framework has been provided so the students can produce a five page report in five days. You may wish to spread the five day project over a weekend or two, to provide more time for reading.

The biggest practical obstacle is a lack of up-to-date print resources. The internet is a vast resource, usually very recent.

People in the field may be the most valuable resource. Nurses, doctors, people treating genetic diseases, users of genetically produced medications, plant and animal breeders, researchers in government and universities, environmental organizations, etc are invaluable. Most communities have some individuals who are involved in this kind of work.

Successful completion of the project will involve a considerable amount of planning on the part of students. Encourage the use of the planning sheet in the Lab Manual. Your role as facilitator should be to refine students' thinking about the focus or scope of their project. Most projects fail because the topic chosen is too broad. The topics mentioned in the activity may be unfamiliar to some students. It may be necessary to spend one period exploring these topics in general, before a student can be certain of their topic.

Science Issues

INTERNET sources are numerous and varied. Many are not suitable because of the depth of treatment. Links from CBC Radio, Scientific American, and Discover Magazine may be more suited to this age group.

Research Project

Reproductive technologies are made possible by the science that underlies them, i.e., genetics and molecular biology. In this project, students will demonstrate their knowledge of that science, and explain its application in the technology.

Biotechnology has contributed to improvements in health care, food production, and life style, but its cost and application have led to a variety of controversial social and ethical issues. These issues should be explored in the project.

The Learning Activity

Students will produce a five page report, plus cover and bibliography. They must choose an *Application of Genetics*, preferably one being used near your community. They must examine:

What, using knowledge from this unit
Where, including facilities needed
Who, including education needed
When did this application begin
Societal implications of this application.

Writing a coherent argument is the greatest student challenge. People tend to assume that "the meaning is in the words," that somehow mere exposure to the words conveys meaning. In such a view, almost any collection of words would be seen as adequate, and if the reader can't figure it out, then... too bad.

Developing a clear, coherent line of thinking out of a well chosen series of sentences is an enormous cultural achievement. For that reason, you should insist upon five written pages, with illustrations adding to this total, not replacing it.

Equipment, Preparation and Resources

Reserve use of the Resource Centre, or Library, and the Computer lab for at least two days. Students will need this time to complete their research. Also, plan to have peer tutors available to help monitor behaviour and to lend assistance during the time that the class is outside of the Science lab.

Prior to this class, put a sample of useful print resources on reserve in the Library. Perhaps this will prevent a common complaint that all the good books are gone.

Categories:	Assessment and Evaluation
Knowledge:	understanding of concepts, terms, relationship between ideas
Inquiry:	technical skills evident in product
Communication:	clarity, precision, knowledge of purpose
Applications, Extensions:	analysis of social and economic issues

Student Exercises

2: Explaining Reproduction

Knowledge and Understanding

Three theories are emphasized in this unit. You are probably already familiar with the cell theory. In addition, you will learn new ideas about cell division. Finally, you will learn how our ideas about DNA allow us to explain how living things reproduce. Additional concepts will be introduced as needed.

We will work with pictures and models to illustrate how reproduction occurs.

Knowledge and understanding are probed at regular intervals in the Grade Nine Daily quizzes. Study these as you go through the exercises, so that you can do your best when they are assigned.

Inquiry and Thinking

As often as possible, we will use the PEOE cycle for most labs and activities. You are expected to frame a question, provide your best prediction, and explain your thinking, using both sentences and diagrams. In other exercises, you will make clones of a plant, a simple worm, and yeast. In other exercises, you will use paper models to predict how genetic damage can occur.

At the end of the unit, you will be given a five day independent project. The project will demonstrate your ability to conduct your own investigation.

Communication

The quality of your arguments is the most important aspect of communication in this chapter. Your arguments consist of sentences, organized into paragraphs, and supported by diagrams or other representations.

Each sentence should be clear and to the point. You will find it best to limit your sentences to two concepts linked together to make a reasonable claim. If you need to relate more than two concepts, add a new sentence.

Applications, Connections and Extensions

Every exercise in this book is designed to support you as you learn appropriate theories and apply them to problems. In the labs, you demonstrate your understanding of a theory only by applying the theory. In the quizzes and projects, you are invited to make further connections and extensions of your learning.

Introduction: Three Theories to Understand Reproduction

In this unit, we will examine cells, the basic structure of life on Earth, and apply the cell theory to processes of cell division and reproduction. This unit consists of three main ideas:

1. **The Cell Theory**   What is the underlying structure of living things? It is widely accepted among scientists that all living things are made of cells. You are made of approximately sixty trillion cells. Your earlobe might have five hundred million cells in it, with perhaps hundreds of distinctly different kinds of cells!   The cell theory involves these four key ideas:

    1. **All living things are composed of cells.** Every living thing that you see is composed of smaller living things: cells. There is still some controversy over this statement, but it is limited to a small number of remarkable forms of life.

    2. **Cells are the basic structural unit of life.** Cells underlie the shapes and structures of all living things that you see. Starfish are different from birds because of the ways that their respective cells adhere to each other.

    3. **Cells are the basic functional unit of life.** All of the functions that keep us alive occur inside our cells. There does not appear to be a "life" that exists outside our cells.

    4. **All cells come from pre-existing cells.** The transmission of life from one generation to the next takes place at the level of the cell.

2. **Cell Division and Reproduction**   All cells appear to have a basic common structure, which permits a tremendous amount of variation. Bone, nerve and blood cells have the same underlying structure, yet have very different properties and roles in the body. We will consider the cellular structures involved in cell division, and how they influence reproduction. Six key ideas about cell division are:

    1. **Multicellular organisms are able to grow and repair tissue because of a process called cytokinesis or cell division.** When you cut your finger, the healing process and subsequent growth involves the division of a very large number of cells.

    2. **Before cells can divide, their nuclei must first divide in *mitosis*.**

    3. **The daughter cells formed by mitosis are genetically identical to the parent cell.**

    4. **Mitosis occurs in almost all body (somatic) cells.**

    5. **Meiosis is a special kind of cell division that produces sperm and eggs cells.**

    6. **Meiosis occurs only in the sex organs.**

3. **DNA and Reproductive Biology** DNA , Deoxy-ribo Nucleic Acid, is the genetic material contained in the cell nucleus. DNA is a long molecule. In human beings, about 1 metre of DNA is twisted up into every nucleus!!

1. **The nucleus of every human cell contains about one metre of DNA.** If one of your cells was the size of a basketball, the total length of DNA inside the cell would be about 300 km.

2. **Every human cell contains 46 pieces of DNA, 23 pieces from each parent.** The pieces are usually all coiled up, and form separate packets.

3. **DNA is a double-stranded molecule with chemical patterns for all cell components.**

4. **Every cell of your body contains the complete DNA pattern for you.** Pull out one hair from your arm. Clinging to the root of the hair are a few cells, each of which contains the entire pattern for your body.

5. **In asexual reproduction (cloning), the child contains an exact copy of one parent's DNA.** We use the words "parent" and "child" here for all organisms, even things like carrots.

6. **In sexual reproduction the child contains equal DNA from both parents.** There are many different ways that different organisms manage to mix DNA from two parents.

All of these ideas are needed to adequately explain how reproduction occurs. You will be required to both memorize the main points, and to apply them to problems in this book.

In all of the exercises in this book, the question must be answered in *complete sentences*. One sentence is one thought. A single word is simply not enough.

Lab 1.1: Using the Microscope

Focus Question: *What are the parts of the microscope? What are their functions?*

1 *Label* the diagram, using your text book

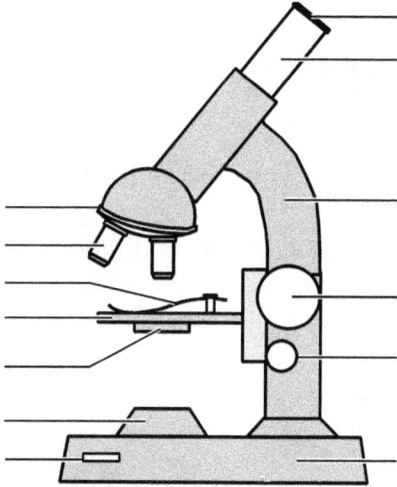

2 *Explain* the function of each part.

Objective: _____

Eyepiece: _____

Stage: _____

Coarse Focus: _____

Fine Focus: _____

Diaphragm: _____

Lamp: _____

3 *Fold* the page along this line, so you can't see the diagram above. Then label it again, with no help from your text or your friends.

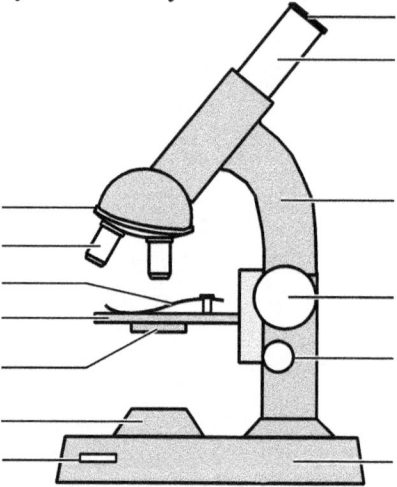

4 *Fold* the page along this line so you can't see the list above. Then list each part, and explain what it does.

_____ : _____

_____ : _____

_____ : _____

_____ : _____

_____ : _____

_____ : _____

_____ : _____

The *Power* of a lens is the number of times the lens magnifies the object. What is the power of your:

Eyepiece: Low Objective: Medium Objective   High Objective:

The *Total Power* is (**Eyepiece** × **Objective**). Find the total power of the:

Low Objective: Medium Objective: High Objective:

Cell Structure and Function

Name:

Date:

Focus Question: How do we prepare a wet mount? How do we focus the microscope?
1.   Study the instructions, and try to memorize them.
2.   Gather together the materials for both exercises.
3.   Give your partner this page.
4.   Your partner will watch and evaluate your progress by checking off each step.

1   Prepare a Wet Mount (You need a microscope slide, cover slip, and millimeter graph paper.)

Holding the slide by the edges, inspect it for cleanliness. Clean if necessary.

Carefully place the millimeter graph paper specimen on the centre of the slide.

Place a drop of water on the specimen.

Holding a clean cover slip by the edges, lower it until one edge makes contact with the slide.

Slowly lower the other edge of the cover slip, so that no air bubbles form.

Use a bit of paper towel to absorb any excess water from around the cover slip.

Turn the slide upside down to show that the cover slip will not fall off.

2   Focus the microscope

Carry the microscope with one hand on the arm, and the other on the base.

Place the microscope gently on the table. Plug the microscope in, and turn on the light.

Turn to Low power. Lower the stage.

Place the prepared slide on the stage, under the stage clips.

While looking from the side, raise the stage as close as possible to the objective, without touching.

While looking through the eyepiece, lower the stage using the coarse focus knob.

Use the fine focus knob to obtain the best image.

3   Put the microscope away.

Turn the objective to lowest power, lower the stage, and remove the slide.

Turn off the light, unplug the cord, and wrap the cord neatly around the microscope.

Use two hands to carry the microscope back to the storage cart.

Which procedures do you need to improve?

Lab 1.2: Drawing from the Microscope

Do you Remember? **List the 4 propositions of the Cell Theory.**

1. _____

2. _____

3. _____

4. _____

What's The Question?

We know that a microscope magnifies things, but how much does it magnify? *How big does one millimetre look under low, medium, or high power?*

What Are We Doing?

1. Prepare a wet mount slide of a small piece of millimetre graph paper.
2. View the millimetre graph paper under low power. Draw an image on the back.
3. Repeat, using medium and high power.

What Are We Thinking About?

1. How big is the field of view under low, medium and high power?
2. Does the brightness of the image appear to change as you switch from low to medium to high power?
3. How thick is one hair? Place one on a slide and determine its thickness.

Your teacher will check your drawings for the following details

1. All lines are drawn in sharp pencil.
2. Clean line drawings, no sketch marks, open circles, careless representations, etc.
3. Your drawing shows at least three levels of detail. Detail inside detail inside object.
4. Your drawing correctly distinguishes between parts of your object.
5. Each view is labelled with the specimen name, and the magnification.

As you make your first microscope drawing, aim for accuracy and speed. Don't erase, and don't use white-out. If you make a mistake, just continue, and make an improvement on your next drawing.

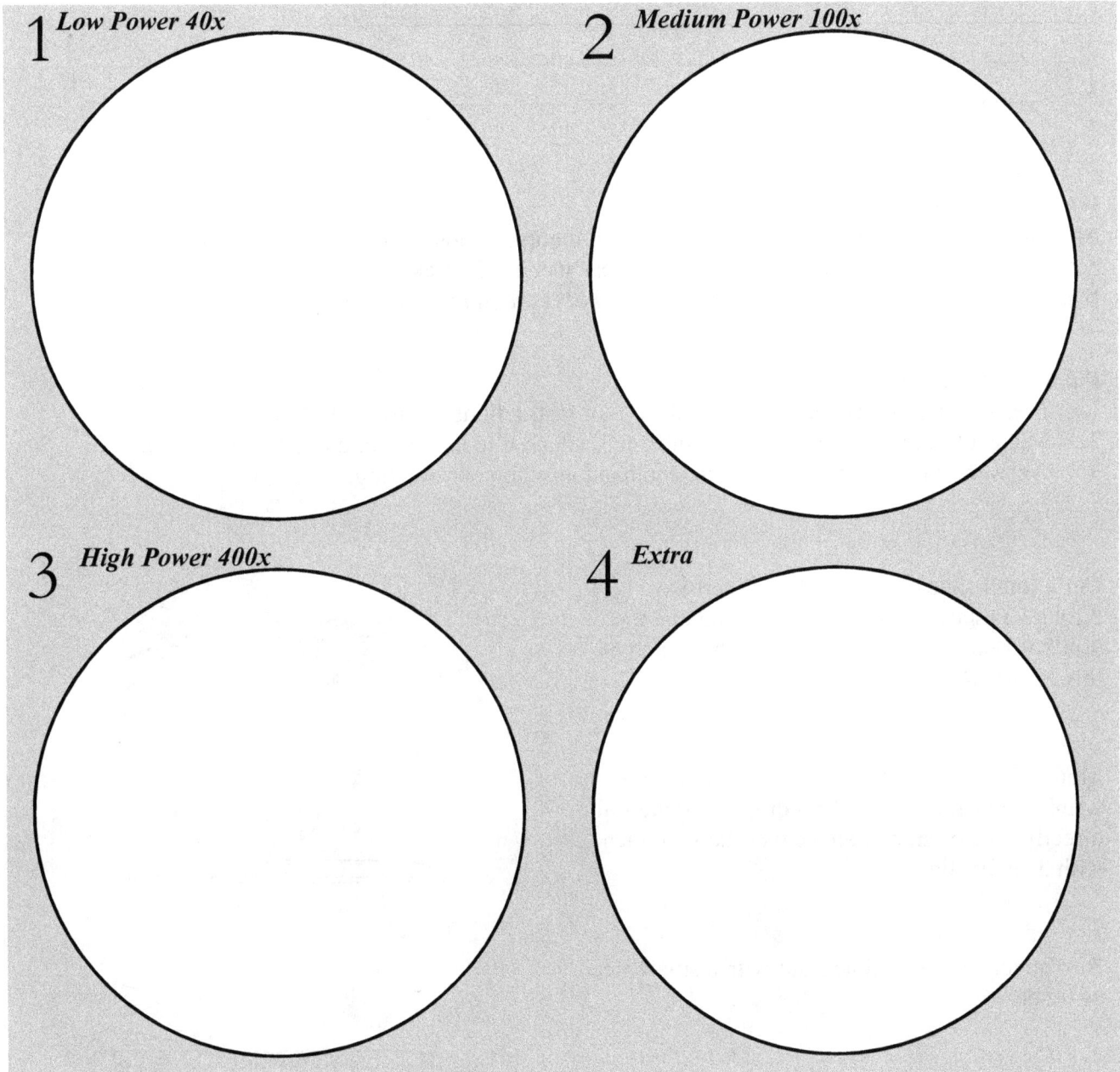

1 *Low Power 40x*

2 *Medium Power 100x*

3 *High Power 400x*

4 *Extra*

Suppose you had a small ball, exactly one tenth of one millimetre across.   The period at the end of this sentence is five times bigger than that. Draw a picture of that ball in each of the diagrams above, at the size that it would appear in your microscope.

How can you use these diagrams to measure the size of all the cells you observe?

Explaining Reproduction

Lab 1.3: The Structure of a Plant Cell

Do you Remember? **List the 4 propositions of the Cell Theory.**

1. _____

2. _____

3. _____

4. _____

What's The Question?

An onion is a living thing. According to the cell theory, it must be made of cells. Onion cells are both large, and easily obtained in thin layers nearly one cell thick.
What is the structure of a typical plant cell? What is the function of each part?

What Are We Doing?

1. Prepare an onion cell slide and view it. Full instructions are found below.
2. Make a full page drawing of the onion cell, attach it to this lab, and hand it in.
3. Answer all questions on this page, and hand in with your drawing.

Cut a small slice of onion, and cut partway through as shown. Break the onion so that the thin layer comes free. Cut a small specimen of this membrane.

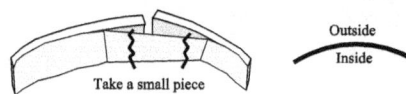

Mount the thin membrane on the slide without wrinkling or folding it. Put a drop of iodine stain directly on the sample, then cover the specimen with a cover slip.

Blot up any excess iodine stain with a small piece of tissue.

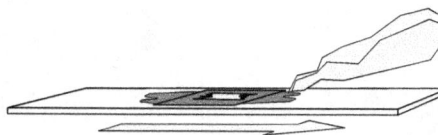

Put a drop of clear water on one side of the cover slip. A piece of paper towel on the opposite side will draw the water through, clearing the slide. Only the onion should remain yellow.

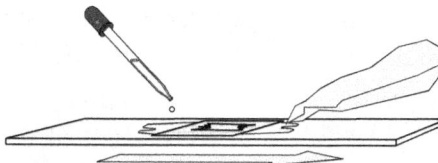

Cell Structure and Function

Name:

Date:

Focus Question: What is the *structure* (parts and their arrangement) of the plant cell? What is the *function* of each part (the jobs they do)?

1 *Use the index of your text to find the parts of the plant cell. These words may help you search your textbook.*

Cell membrane   Nucleus
Nucleolus   Cell Wall
DNA   Vacuole
Cytoplasm   Mitochondrion
Ribosomes   Lysosome
Nuclear Membrane   Chloroplasts

2 *Use your textbook to label the plant cell. Find and label as many of the words on the left as you can.*

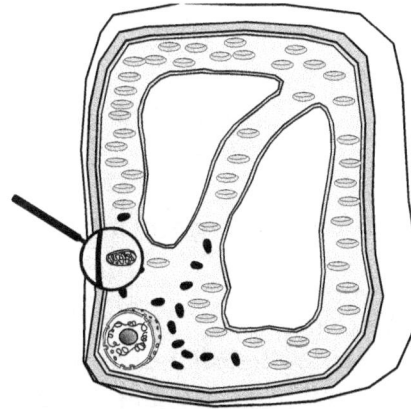

3 *Write a brief description of the function of each of these parts.*

Cell Membrane

Nucleus

Cell Wall

Vacuole

Chloroplasts

Mitochondrion

Lab 1.4: The Structure of an Animal Cell

Do you Remember? **List the 4 propositions of the Cell Theory.**

1. _____

2. _____

3. _____

4. _____

What's The Question?

According to the cell theory of life, we must be made of cells. These cells must carry out all of the functions that make us alive.

What is the structure of a typical animal cell? What is the function of each part?

What Are We Doing?

1. Prepare a cheek cell slide and view it. Full instructions are found below.
2. Make two full page drawings of the cheek cell, attach them to this lab, and hand them in.
3. Answer all questions on this page, and hand in with your drawings.

Obtain some cheek cells by scraping the inside of your cheek with a toothpick. A tiny amount of white tissue should come free. Smear the cells on the slide, and place a cover slip on the slide.

Place one drop of methylene blue stain on one side of the cover slip. With a paper towel on the other side of the cover slip, draw stain through the wet mount. Blot up any excess stain. Careful: this stuff really stains clothes and skin.

Clear the slide as before. The cells should look like tiny blue specks on a clear slide, with no excess blue stain.

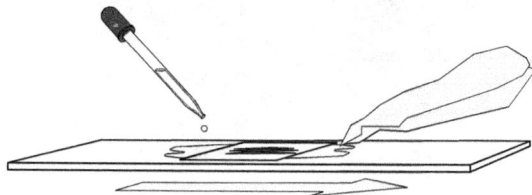

Cell Structure and Function

Name:

Date:

Focus Question: What is the *structure* (parts and their arrangement) of the animal cell? What is the *function* of each part (the jobs they do)?

1 *Use the index of your text to find the parts of the animal cell. These words may help you search your textbook.*

Cell membrane   Nucleus
Nucleolus   Centriole
DNA   Vacuole
Cytoplasm   Mitochondrion
Ribosomes   Lysosomes
Nuclear Membrane

2 *Use your textbook to label the animal cell. Find and label as many of the words on the left as you can.*

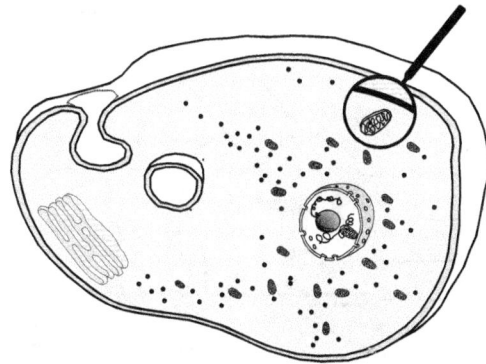

3 *Write a brief description of the function of each of these parts*

Cell Membrane

Nucleus

DNA

Vacuole

Cytoplasm

Mitochondrion

The Grade Nine Daily

Quiz 1.5: Cell Structure and Function

1 Label the following parts of the Microscope

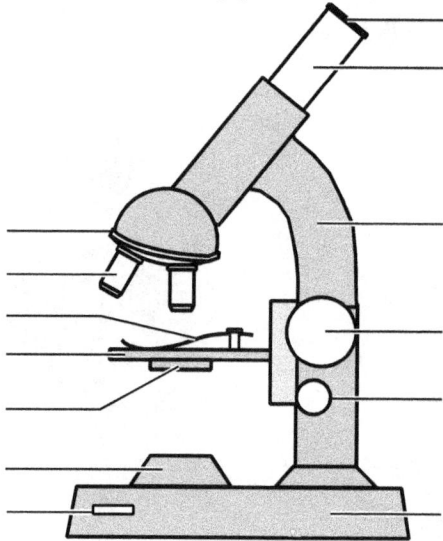

Date: _____   / 4

2 Name the parts of the microscope that perform these functions.

a) _____ you look through this
b) _____ supports the microscope
c) _____ holds slides down
d) _____ moves slide closer and farther
e) _____ allows you to switch powers
f) _____ lens closest to the slide
g) _____ adjusts brightness of light
h) _____ produces light
i) _____ moves slide up quickly
j) _____ moves slide up slowly

Date: _____   / 4

3 Estimate length of one cell in each view.

A 40x Φ = 4000 µm **B 100x Φ = 1600 µm**

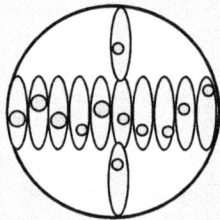 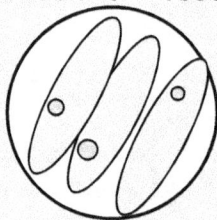

C 400x Φ = 400 µm **D 400x Φ = 400 µm**

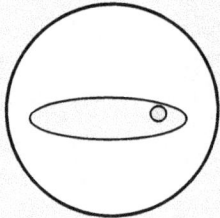 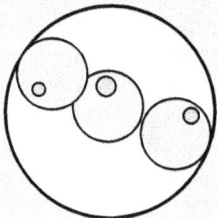

Date: _____   / 4

4 This is a 100× view. Estimate the length of:

_____plant cell   _____animal cell

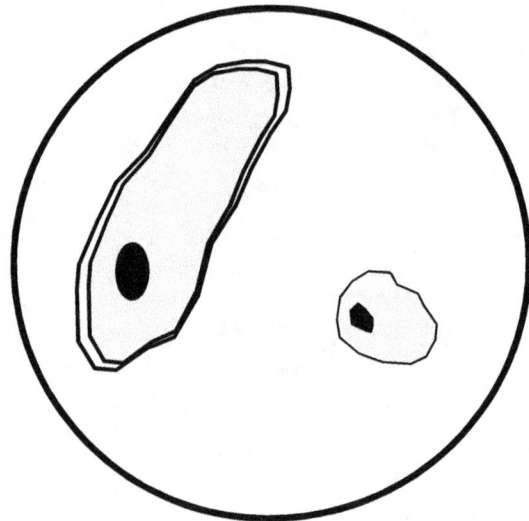

Date: _____   / 4

Quiz 1.5: **Cell Structure and Function**   **Name:**

5   List 4 points of the Cell Theory

1. _____

_____

2. _____

_____

3. _____

_____

4. _____

_____

Date: _____   / 4

6   Use a sharp pencil to *outline* a cell and its nucleus in the following diagram. *Label* those parts that are visible.

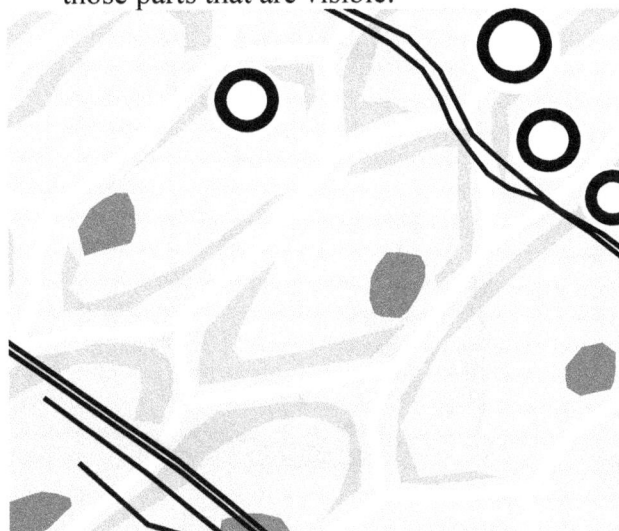

Date: _____   / 4

7   Label the following parts of the microscope.

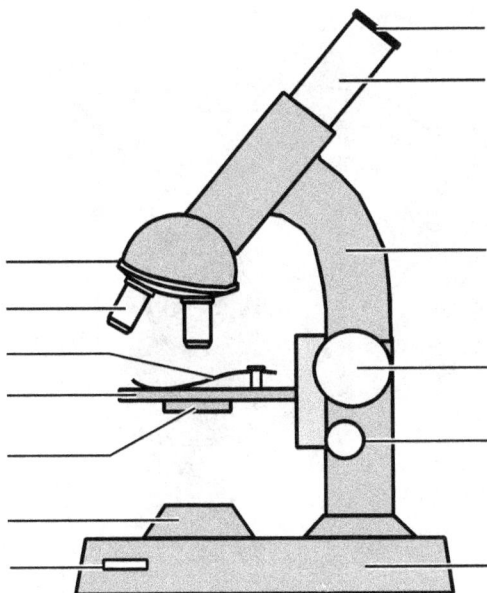

Date: _____   / 4

8   Name the parts of the microscope that perform these functions.

a) _____ you look through this

b) _____ supports the microscope

c) _____ holds slides down

d) _____ moves slide closer and farther

e) _____ allows you to switch powers

f) _____ lens closest to the slide

g) _____ adjusts brightness of light

h) _____ produces light

i) _____ moves slide up quickly

j) _____ moves slide up slowly.

Date: _____   / 4

All the news that's fit to print...   and then some

The Grade Nine Daily

Quiz 1.5:   Cell Structure and Function   Name:

9 Label the nucleus, cell wall, cytoplasm, vacuole, chloroplast, nuclear membrane, cell membrane, DNA, and nucleolus.

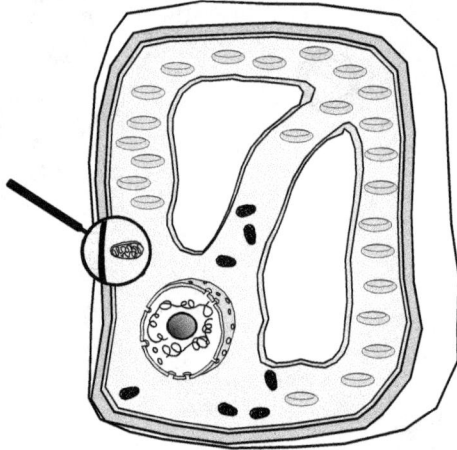

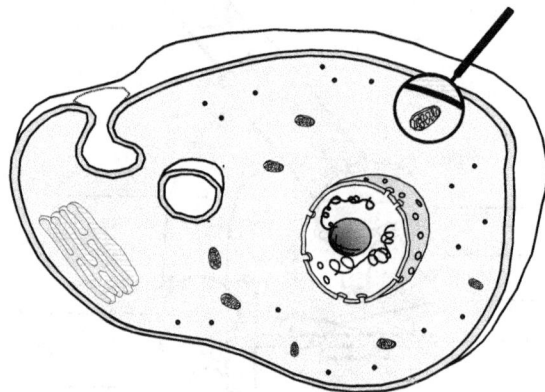

Date: _____   / 4

10 What part of the animal cell:

1. _____ breaks down food (respiration)?
2. _____ is inherited from the parent?
3. _____ guides chromosomes in mitosis?
4. _____ is copied during reproduction?
5. _____ help instructions move to cell?
6. _____ are the instructions for the cell?
7. _____ selects what enters cell?
8. _____ is the "living stuff" of the cell?
9. _____ stores water, minerals & wastes?
10. _____ controls all cell processes?

Date: _____   / 4

11 List the main points of the Cell Theory.

1. _____
   _____
2. _____
   _____
3. _____
   _____
4. _____
   _____
5. _____
   _____

Date: _____   / 4

12 Identify the nucleus, cell wall, cytoplasm, vacuole, chloroplast, nuclear membrane, Cell membrane, DNA, and nucleolus.

Date: _____   / 4

r__

© *Ross Lattner Publishing*   **50**   *www.rosslattner.ca*

13   What part of the cell contains the instructions for all activities within the cell? Explain.

Date: / 4

14   When a cell grows, does it get more molecules, or does it just get bigger? Explain.

Date: / 4

15   *"DNA is a molecule that contains instructions for the cell".*
How can a molecule contain instructions?

Date: / 4

16   Are the *instructions* for the cell's activities made out of matter, or out of something else? Write a paragraph to explain your answer.

Date: / 4

Activity 2.1: Cell Growth and Reproduction.

A student has drawn eight diagrams of 8 different cells at different stages of their growth. Each card, labelled A to H, shows a cell at a particular stage of growth. These diagrams are found on page 53.

1. Arrange the cards in the correct order, on the grid on the following page.
2. In the table below, write the letter of each card in the order you have chosen.

Student Names	1	2	3	4	5	6	7	8

3. Copy the letters of two other groups into the table above. Are they in the same order? In what ways are they the same? In what ways are they different?
4. After the class discussion, glue the cards into the grid on the following page, in the correct order.
5. Label all the cell parts in card number 4 on the back of this sheet.
6. Look up these words in your text book or use the Internet. Above each card, write one of the following words: *Cell Growth*   *Mitosis*   *Cytokinesis.*

7. When cells are growing, do they get *more particles*? Or do they just get *more volume*? Explain.

8. Write the story of the life cycle of a cell in one paragraph.

Cell Growth and Reproduction

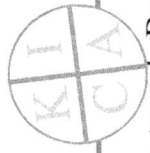

Name:

Date:

Activity: 2.1: Cell Growth and Reproduction: Arrange the cards on this diagram in the order that progresses smoothly from beginning to end. Do not glue or draw the cards until after the class discussion.

1	2	3	4
5	6	7	8

53

www.rosslattner.ca

© Ross Lattner Inc.

Explaining Reproduction

Cards for
Activity 2.1:

Cell Growth and
Reproduction.

See page 51.

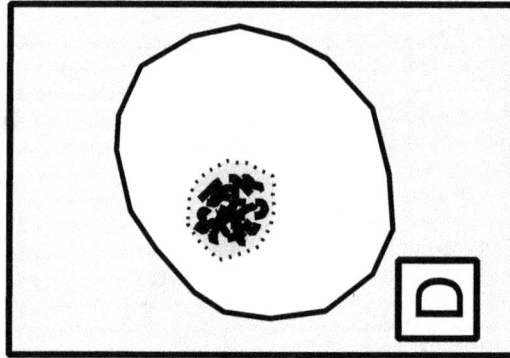

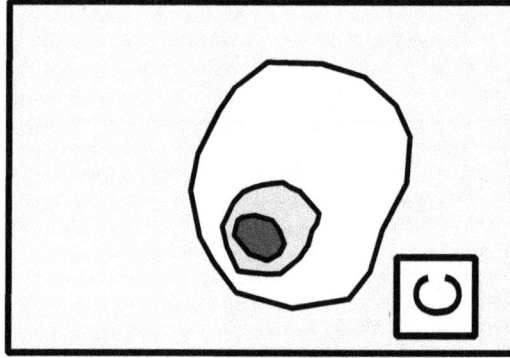

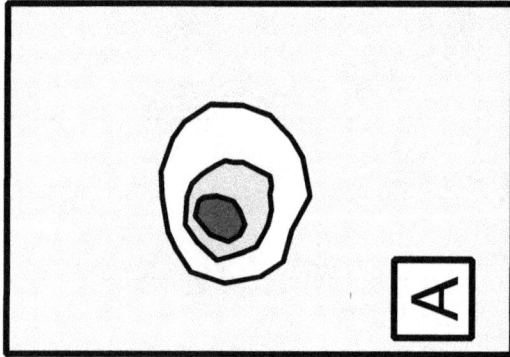

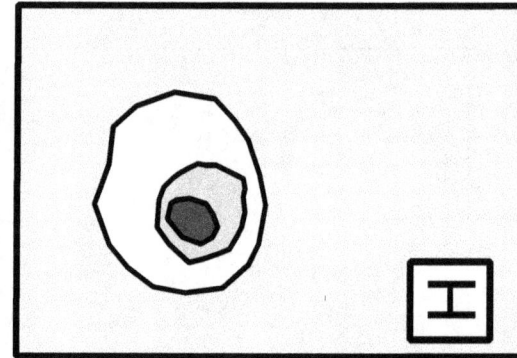

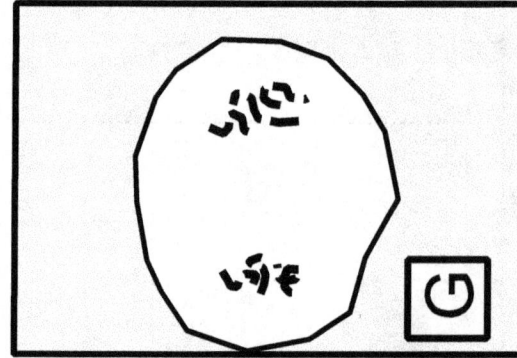

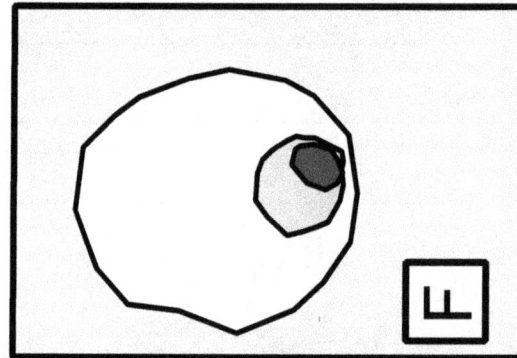

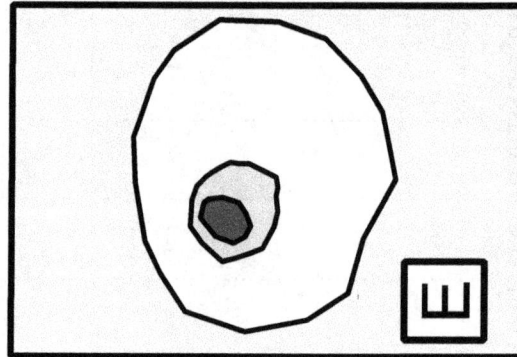

wsw.rosslattner.ca

Cell Growth and Reproduction

Activity 2.2: Nuclear Reproduction (Mitosis) and Cell Division

A student has drawn eight diagrams of eight different cells at different stages of mitosis. Each card, labelled P to W, shows a cell at a particular stage of mitosis. These diagrams are found on page 56.

1. Arrange the cards in the correct order, on the grid on the following page.
2. In the table below, write the letter of each card in the order you have chosen.

Student Names	1	2	3	4	5	6	7	8
1								
2								
3								
Group into Stages of Mitosis:								

3. Go to two other work groups. Copy their names and their letters into your table. Are their cards in the same order? In what ways are your lists the same? In what ways are they different?
4. Read your text book account of *mitosis*. In the table above, label the stages of mitosis.
5. After the class discussion, draw or glue the cards into the grid on the following page, in their correct order. Name the stages of mitosis above each card.
6. Write a brief description of each stage of mitosis in the spaces below. Pay particular attention to the roles and changes taking place in the *nuclear membrane, DNA, centrioles, chromosomes, chromatids, centromeres,* and *spindle fibres.*

Prophase

Metaphase

Anaphase

Telophase

Cytokinesis

Cell Growth and Reproduction

Name:

Date:

Activity 2.2: Nuclear Reproduction (Mitosis) and Cell Division

Arrange the cards on this table so that the diagrams progress smoothly from beginning to end. Don't glue the cards until after the class discussion.

1	2	3	4
5	6	7	8

© Ross Lattner Inc.

56

www.rosslattner.ca

Explaining Reproduction

9 Academic Science Lab Manual

Cards for Activity 2.2:

N u c l e a r Reproduction (Mitosis) and Cell Division

See page 54.

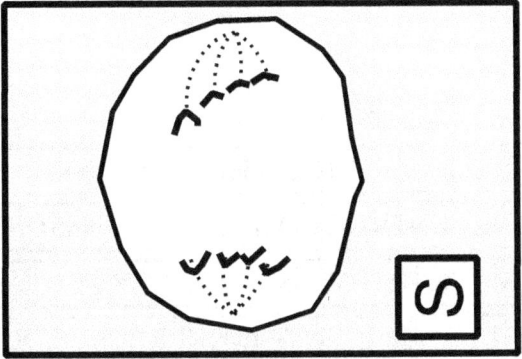

P

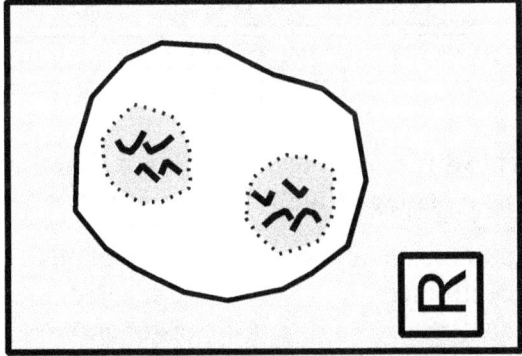

Q

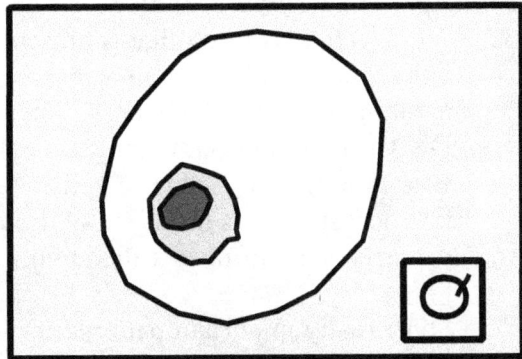

R

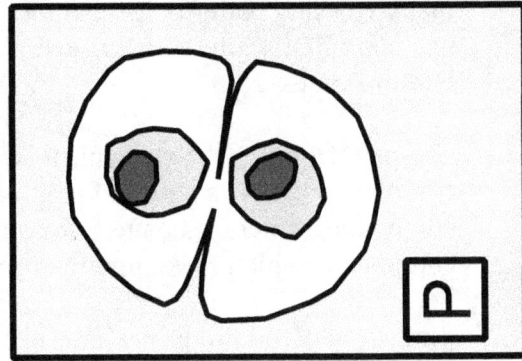

S

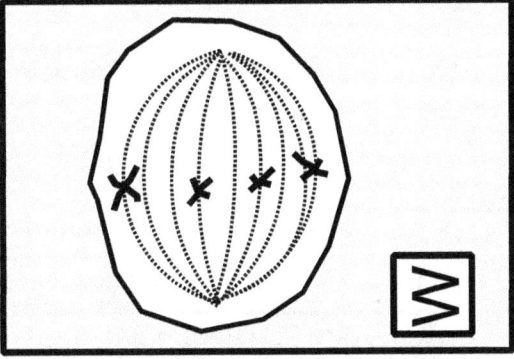

T

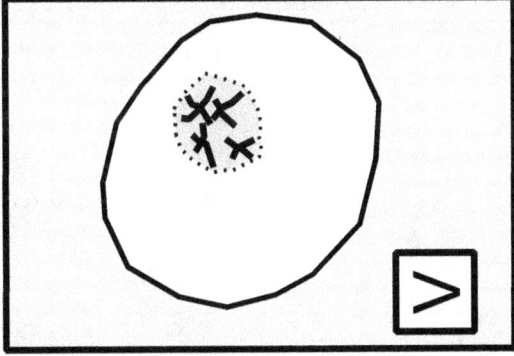

U

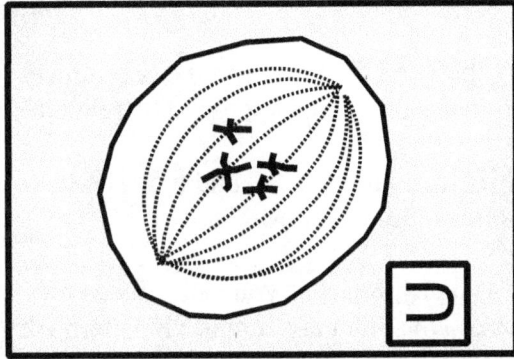

V

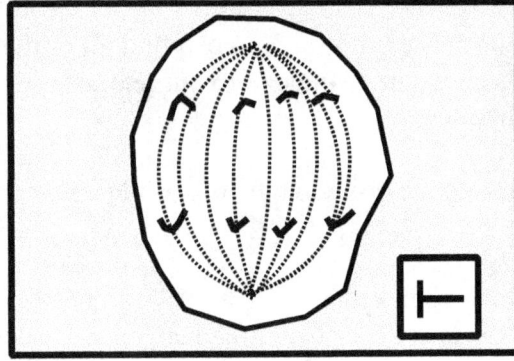

W

Activity 3.1: DNA, Genes and Chromosomes

Imagine Romeo, a model boy with only six chromosomes.

XXMXJUNKCOLOURSIGHTJUNKBLOODAJUNKCENTRJUNKESTROGENJUNKXXXX
MUNKNKNKCURLYHAIRJUNKJUNKTALLJUNKCENTRJUNKMIGRAINEJUNKJUNK
YPYYNOTHINGJUNKUNKBLOODOJUNKJUNKCENTRTESTOSTERONEYYYY
MUNKJUNKNEARSIGHTEDJUNKJUNKJUNKBLUEEYESCENTRJUNKFRECKLESJUNK
PUNKSTRAIGHTHAIRKUNJJUNKSHORTJUNKCENTRHEALTHYBRAINJUNKJUNK
PUNKJUNKFARSIGHTEDJJUUNNKKJUNKBROWNEYESJUNKCENTRNOFRECKLESJNK

What's The Question? We have heard a lot of different words about genetics. *What do these words mean?* Use your text book and this exercise to learn the meaning of the following terms:

gene	allele	chromosome	homologous
diploid	haploid	centromere	replicated
unreplicated	condensed chromosome	chromatid	sister chromatids

What Are We Doing?

1. **Study** the strands of DNA in the box above. Which pairs of strands appear to be related?

2. **Identify** the "junk DNA," and mark it all with one colour of high-lighter.

3. **Find** as many genes as you can. Genes are pieces of DNA that are actually a pattern for something. Mark them with other colours.

4. **Study** your text book. You can find all of the words above in the index of your text, or use the Internet.

5. **Complete** the boxes on the opposite page.

What Are We Thinking About?

1. "Junk DNA" is DNA that is present in your body, but doesn't seem to be a pattern for anything that your body uses. The junk DNA is mutating right along with all of your other DNA. Some scientists believe that occasionally a bit of really useful pattern results from so-called junk DNA.

2. Some really important patterns are repeated many times. Your DNA contains many working patterns (genes) for haemoglobin, scattered over several chromosomes.

3. The path from DNA to organism is very complex. DNA is a pattern for another compound, which is a pattern for other compounds which make up organisms.

Questions For Later...

1. You know yourself better than anyone else does. Which genes do you think you have expressed from your mother? Which from your father?

2. In this exercise, each gene had two alleles. Must there *always* be two alleles in a given individual like Romeo? Explain your thinking.

Genes, Chromosomes and DNA

Name:

Date:

Focus Question: Write the question that you are trying to answer.

1 ***Identify the homologous pairs of chromosomes.*** List the first 10 letters of each pair, one immediately above the other. Are the homologous pairs identical? In what ways are they similar? In what ways different?

2 ***Identify all of the pairs of alleles***

3 ***Find the centromeres.*** Are they in the same location on homologous chromosomes?

4 ***Is the "junk DNA" the same on homologous chromosomes?*** Explain.

Explaining Reproduction

Activity 3.2: DNA During Mitosis

Eve is a model girl, whose DNA is similar to Romeo's (they are both model humans), but is different in many ways.

XXXPUNKCOLOURSIGHTJUNKGBLOODBJUNKCENTRJUNKESTROGENJUNKXXXX
PUNKNKNKWAVYHAIRJUNKJUNKMEDIUMJUNKCENTRJUNKHEALTHYBRAINJNK
XXMXCOLOURBLINDJUNKJUNKBLOODOJUNKJUNKCENTRJUNKESTROGENXXX
MUNKJUNKRIGHTSIGHTEDJUNKJUNKJUNKGREENEYESCENTRJUNKFRECKLESJNK
MUNKSTRAIGHTHAIRKUNJJUNKTALLJUNKCENTRHEALTHYBRAINJUNKJUNK
JUNKPUNKNEARSIGHTEDJJUUNNKKJUNKGREYEYESJUNKCENTRNOFRECKLESJNK

What's The Question? *How is the genetic pattern for a cell transmitted faithfully from one cell to the next? How long does it take for one parent cell to become two daughter cells?*

What Are We Doing?

1. Your group will make six models of a cell at each stage of mitosis and cell division, and fasten them all to a single sheet of Bristol board to make a poster.
2. Six strands of DNA can be found in the box above. Your teacher will have photocopies of these strands. Carefully cut out each strip.
3. Draw a cell membrane with a nuclear membrane inside of this on a piece of paper using a marker.
4. Proceed through each step of the cell cycle, making a new model for each step.
5. At each stage, answer the questions in the space provided and make a diagram of the cell separately.
6. Repeat the cycle several times in order to familiarize yourself with each part of the process.

What Are We Thinking About?

1. What does *diploid* mean?
2. *Homo*: the same; *Logos*: meaning. Hence, *homologous chromosomes* are pairs of chromosomes that have the same meaning.
3. What is the difference between *mitosis* and *cytokinesis*?
4. What is the difference between *mitosis* and *meiosis*?
5. How many chromosomes are present in the cells of your mythical creature before and after one cell cycle?
6. How do the chromosomes compare before and after one cell cycle?

Questions For Later... (to be answered on a separate page)

1. Do all cells of the body contain the same genetic information?

2. Do all cells of the body look like one another? Do they perform the same jobs?

3. How does a human being grow from a single fertilized cell into an individual containing trillions of cells, with different appearances and different jobs?

Genes, Chromosomes and DNA

Name:

Date:

1. Interphase and Chromosome Replication

Throughout *interphase*, the chromosomes are not visible in a light microscope. The DNA is uncoiled, so it can be read inside the nucleus. Read the DNA strands, and then follow these instructions:

A   Identify the *genes*, and colour them. Make all of the genes for height the same colour. Different instructions for the same characteristic are called *alleles*. Genes for human eye colour can be one of four alleles: blue, brown, hazel, green.

B   Group the six strands into pairs. The pairs contain patterns for similar kinds of things. These pairs are called *homologous pairs*. Each homologous pair ought to have similar colours.

C   Replicate each DNA strands in Eve's cell by faithfully copying each letter onto new strips of paper. Keep each replicated pair together , with a paper clip at the *centromere*.

Questions (*to be answered in your notebook*)

1.   What does *diploid* mean? Are most human cells diploid?

2.   How many *alleles* exist for each gene within our mythical person?

3.   How many pairs of *homologous DNA strands* are present in your mythical cell?

4.   Are the homologous DNA strands attached to one another or are they independent in the cell?

5.   What is a chromatid made of (protein, carbohydrate, lipid, and/or DNA)?

2. Prophase of Mitosis

This is the first stage of mitosis. The replicated DNA strands condense into thick coils and become visible. Coil up the strands of DNA into *chromosomes*, leaving them joined at the centromeres. The rest of your model cell does not change. Each pair of sister *chromatids*, linked at the centromere, is called a *replicated chromosome* (or in many books, simply chromosomes).

Questions (*to be answered in your notebook*)

6.   Are the two sister chromatids that are connected by a centromere identical to one another or do they contain different genetic information?

7.   A diploid human cell contains 46 unreplicated chromosomes in early interphase. How many sister chromatids will be present in the human cell during prophase of mitosis?

8.   Can you read the coiled up DNA? Can the cell "read" the coiled up DNA?

9.   Can you make a copy of DNA when it is all coiled up?

10.   Why is it important that mitosis is a brief process in the life history of a cell?

3. Metaphase of Mitosis

In metaphase, the nuclear membrane disappears. *Spindle fibres* form, coming from two structures called *centrioles* that have migrated to opposite poles of the cell.

Draw centrioles on your paper at opposite ends of the cell. Use tape to connect thread (spindle fibres) from the centrioles to the paperclip centromeres. Since the spindle fibres are pulling each set of replicated cromosomes in opposite directions, all of the chromosomes line up across the middle of the cell (metaphase plane). Note: homologous chromosomes are independent of one another.

Questions (*to be answered in your notebook*)

11. How many replicated chromosomes are on the metaphase plane of your cell?

12. How many replicated chromosomes would be on the metaphase plane of a human cell undergoing mitosis?

4. Anaphase of Mitosis

In anaphase, sister chromatids are pulled apart by the spindle fibres to become daughter chromosomes.

Separate your coiled paper daughter chromosomes to opposite ends of the cell. In living cells, chromatids are flexible. They bend on each side of the centromere as they are dragged through the cytoplasm by the shortening spindle fibres.

Questions (*to be answered in your notebook*)

13. Are the two sets of daughter chromosomes moving to opposite sides of the cell identical to each other?

14. Are the two sets of daughter chromosomes identical to the parent cell?

15. Why is this process important for the function of life?

5. *Telophase of Mitosis*

Daughter chromosomes reach opposite poles of the cells. The spindle fibres disappear. Finally, the nuclear membrane begins to form around the chromosomes in each nucleus.

Remove your spindle fibres and draw two new nuclear membranes around the new nuclei. Also, pinch in the line representing the cell membrane.

Questions (*to be answered in your notebook*)

16. Use your text book or the Internet to try to determine what cellular structures make up the spindle fibres.

17. Does a nuclear membrane separate the genetic material from the cytoplasm in all types of cells?

6. *Cytokinesis*

Cyto: of life; *kine*: to cut. Thus, *cytokinesis* is the cutting of the stuff of life into two parts, that is, two new cells. In animal cells, the cell membrane pinches at the centre, from the outside in. The two daughter cells become surrounded by an intact plasma membrane and enter early interphase.

Divide your cell in half in this manner by drawing an increasingly pinched centre that finally touches. Where it touches, it joins to make two roughly spherical cell bodies.

Questions (*to be answered in your notebook*)

18. Does the parent cell still exist?

19. How are the two daughter cells related to each other?

20. How are the daughter cells related to the parent cell?

21. You have used materials to make a model of a cell undergoing mitosis (nuclear division) and cellular division. Explain some ways in which your model differs from the actual process of cell division in animals

Lab 3.3: Observing Mitosis in Prepared Slides

Do you Remember? **List and describe the four stages of mitosis (nuclear division)**

1. _____

2. _____

3. _____

4. _____

What's The Question?
Mitosis is the process by which one nucleus makes an identical copy of itself. *What does the process of mitosis look like in a prepared slide?*

What Are We Doing?

1. Place a prepared slide under medium power, and look for cells with visible chromosomes.

2. Change to high power, and draw the cells in the table on the back.

3. Repeat until you have observed and drawn four stages of mitosis.

What Are We Thinking About?

1. Not all parts of mitosis are visible in every kind of cell.

2. You need to observe several cells at the same stage to get the whole picture.

3. Do you remember how to draw from a microscope? You can review the skills you covered in Lab 3.2 if you have forgotten.

Questions For Later...

1. Were all of the cells undergoing mitosis in the slides which you observed?

2. Which stage of mitosis seems to be hardest to find? Which is easiest to find?

3. From (2), what can you infer about which stage of mitosis takes the longest time to complete? Which stage takes the shortest time to complete?

1. After one cell cycle, two daughter cells are formed. After 2 cell cycles, four daughter cells are present. How many cell cycles (cell divisions) are necessary to produce 100 cells?

Genes, Chromosomes and DNA

Name:

Date:

Make one complete drawing for each stage. Make them as large as possible within the space below. Capture details from several cells to make one complete composite with clear features. Be sure that your drawing has clean pencil lines, no sketch marks, no careless representations, and is fully labeled.

Prophase

Metaphase

Anaphase

Telophase

Lab 3.4: DNA and How It Works

What's The Question? DNA (deoxyribonucleic acid) is the genetic material found in all cells. It contains all of the patterns to make new cells and to control cell activities. It is passed from parent to offspring during reproduction.
What does DNA look like, and how is it able to make new copies of itself?

1 *Make the Chain* Use the accompanying paper DNA template to make a chain of 10 nucleotides. Cut and paste the nucleotides on a piece of blank paper so that the **P**hosphate on one is bonded to the **D**eoxyribose of the next one.

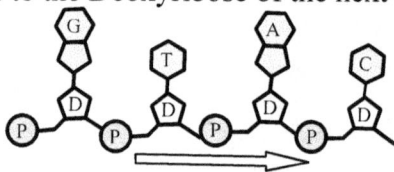

Any sequence will do, but you will not be able to compare your results with others in the next two lab exercises. If you wish to compare your work to that of others, make your sequence:
adenine, guanine, guanine, cytosine, cytosine, adenine, thymine, adenine, adenine, guanine .

2. *Make the complementary strand.* Besides making a P-D-P-D chain, DNA can bond across as well. However, only two combinations bond together. **A**denine and **T**hymine can bond together; **C**ytosine and **G**uanine can bond. When this happens, you get a second string of DNA sticking to the first:

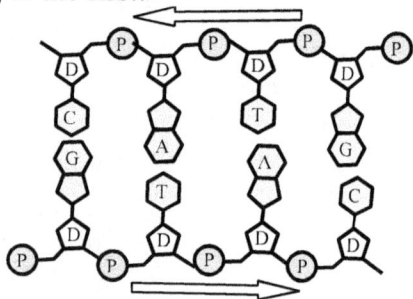

This "ladder" actually twists into the famous double helix shape of DNA.

3. *Unzip the complementary chain into two.* DNA is able to unzip into two strands, by separating right down the middle, breaking the A-T bonds and the C-G bonds.

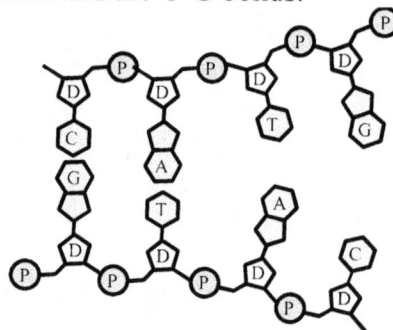

4. *Rebuild two new strands* by repeating Step 2 on both of the single strands.

5. *Compare the two new double strands* to see if indeed they are the same.

Now think about the process that must be occurring inside your body. Remember that your temperature is 37°C. Water molecules will be moving at about the speed of sound at that temperature, or about as fast as a rifle bullet.
Imagine this process taking place in the Skydome. Traveling at the speed of sound, baseball-sized molecules of water pound a strand of DNA one million baseballs long. Is it not remarkable that a nearly perfect copy can be made under those conditions?
On the other hand, at 37°C the whole process occurs with impressive speed, often in a matter of minutes!!

Genes, Chromosomes and DNA

Questions For Later...

1. What is the sequence of your complimentary strand of DNA?

2. Is it possible for you to make a mistake in the sequence mentioned above (assuming you always match adenine to thymine, and guanine to cytosine)?

3. When does the process of replication occur in the cell's life cycle?

4. Suggest some ways that DNA replication could be made inaccurate.

Explaining Reproduction

Lab 3.5: Genes and How They Work

List the four nucleotides and its complementary partner (e.g., adenine pairs with ....)

1. _____

2. _____

3. _____

4. _____

What's The Question?

Genes are sequences of nucleotides found in our DNA. Each gene is a pattern for a protein. The proteins function mostly as catalysts which speed up important biological reactions. Genetic traits are expressed as a result of the activities of proteins patterned after the DNA.

How does DNA serve as a pattern for proteins? What happens if the pattern mutates?

What Are We Doing?

1. Make up a 30 nucleotide sequence of DNA by writing the first letters of the four nucleotides (e.g., adenine - A, cytosine - C, and so on) in any order you decide.

2. Starting with the first three letters (which are called codons) and every three letters after that, use the table of the genetic code to prepare a list of the 10 amino acids specified by your DNA. See page 84 for the code!

3. Cut out the appropriate amino acids from the sheet supplied by your teacher.

4. Tape the sequence of 10 amino acids together to make a chain.

5. One at a time, apply the three major sources of mutational damage to your DNA sequence. Determine the effect on the DNA (if any) by repeating steps 2 - 4 to make three new strands of proteins.

6. Compare the effects of the three mutational agents on the resulting sequence of amino acids.

Questions For Later...

1. Did every mutation result in a change in the sequence of amino acids produced?

2. Is there any punctuation in the genetic code (e.g., is there a space between codons)?

3. Would a mutation that deleted three nucleotides be more or less serious than a mutation that deleted only one or two nucleotides? Explain your answer.

4. What can we do to reduce our risk of mutations?

Genes, Chromosomes and DNA

Name:

Date:

Focus Question: Using the letters C G A and T, write a random sequence of 30 nucleotides in this box . This will be your initial DNA sequence.

---

1 **Initial Protein (a sequence of 10 amino acids)**

List your amino acids in order.

2 **Mutagen: High Energy Radiation**
Effect of mutagen: *X*-ray strikes initial DNA sequence and causes random deletion of a single nucleotide.
List your amino acids in new order.

---

3 **Mutagen: Ultra - Violet Light**

Effect of mutagen: affects only adjacent thymine nucleotide paris. This prevents any further addition of amino acids The rest of the DNA is simply not read.
List your amino acids in new order.

4 **Mutagen: Chemical Agents.**

Effect of mutagen: changes all cytosine (C) to thymine (T) in initial DNA sequence.
List your amino acids in new order.

Check the Genetic Code on Page 84 to find how to translate the DNA code into amino acids!

Nucleotides to Build DNA models

Adenine

Guanine

Cytosine

Thymine

Name:

Date:

Eighty Amino Acids

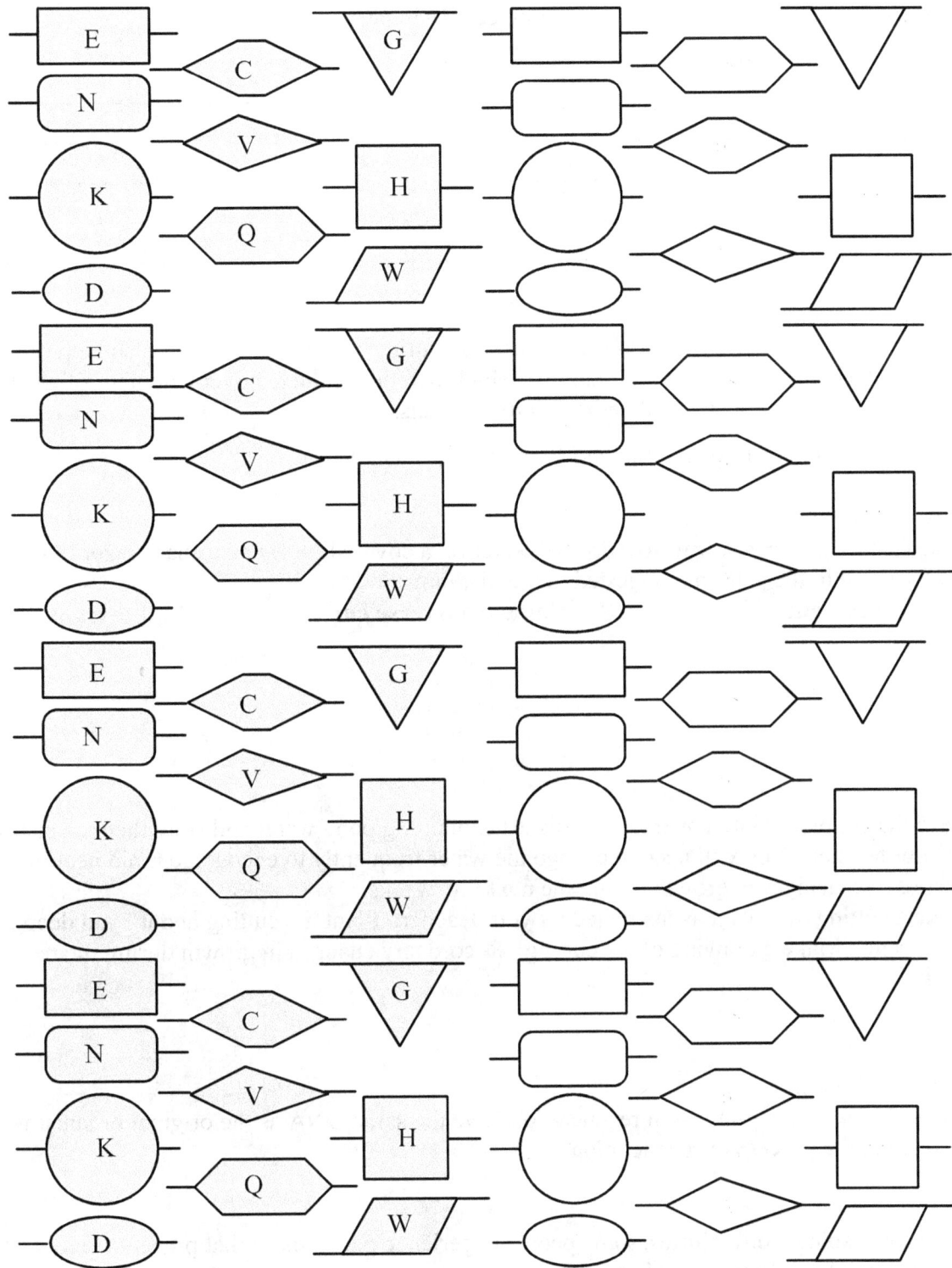

Your teacher will have extra copies of this page. Cut these out to make your amino acid chains.

Lab 4.1: Asexual Reproduction in Plants and Animals

Do you Remember? List the six key ideas about cell division.

1. _____
2. _____
3. _____
4. _____
5. _____
6. _____

What's The Question?

It is possible to grow a whole plant from a leaf or root cutting. Some animals can also be reproduced this way! How is this possible? In this activity, which will be conducted over a number of weeks, you will attempt to regenerate plants and perhaps some animals.

How is the genetic information transferred from one organism to another?

What Are We Doing?

1.   Transfer the *Planaria* worms to a drop of water on a cover slip. Using a sharp razor blade, cut the *Planaria* in the patterns suggested in the diagrams:

Planaria head cut   **Single transverse cut**   **Double transverse cut**

2.   Place the cut pieces into covered petri dishes containing pond water and store them as instructed by your teacher. You will need to change the water frequently to encourage rapid healing. Record any changes in growth during the next few weeks.
3.   Make a cutting of a plant as instructed by your teacher. Plant the cutting about 3 cm deep in moist sand. Make a drawing of the cutting. Record any changes in growth during the next few weeks.

Questions For Later...

1.   Do the ***Planaria*** resulting from regeneration have the same DNA as the original organism or is it changed in the process of regeneration?

2.   After some spinal cord injuries, some people experience paralysis. What prevents regeneration of nerve tissue to allow this to happen?

Varieties of Reproduction

Name:

Date:

K | I
C | A

Focus Question: Write the question that you are trying to answer.

1 **Predict:** What will happen to the plant and animal after they are cut and placed in a growth medium?

2 **Explain:** Why will this happen?

3 **Observe,** and record your observations here.

4 **Explain** your observations.

Lab 4.2: Viewing Asexual Reproduction in Prepared Slides

Do you Remember? List the 6 key ideas about cell division.

1. _____
2. _____
3. _____
4. _____
5. _____
6. _____

What's The Question?

Certain organisms (e.g., yeast and bread mold) can reproduce without need of any partner? What does that look like? In this lab you will use the microscope to make drawings of asexual reproduction in some primitive organisms.

What Are We Doing?

1. Working in pairs, obtain a slide from your teacher and a microscope.

2. Bring the slide into focus at low power, then search for signs of budding, or sporulation. Move to medium, and high power, if possible. Make a drawing of the slide to show the distribution of cytoplasm in the new cells in the space provided. Show at least two levels of detail in your drawing.

3. If other slides are available, make drawings of these as well.

What Are We Thinking About?

1. What is the genetic relationship between different spores in the sporangium of a bread mold?

2. In this diagram of yeast budding, there is not an equal distribution of cytoplasm. Would the bud contain less DNA than the "mother"?

Questions For Later...

1. What conditions are necessary for molds to form on bread?

2. Identify three ways that fungi can reproduce asexually.

3. Yeast is used to make bread rise. How does this happen? What conditions are necessary and what causes the great smell that is evident when bread is rising?

Varieties of Reproduction

Name:

Date:

Focus Question: Write the question that you are trying to answer.

1 *Budding in yeast* (draw your diagram in this space)

2 *Fission in bacteria or protozoans* (draw your diagram in this space)

3 *Sporulation in fungi* (draw your diagram here)

4 *Explain* the differences among budding in fungi, fission in bacteria or protozoa, and sporulation in fungi.

Explaining Reproduction

Lab 4.3: Sexual Reproduction in Animals

Compare Mitosis with Meiosis using the following categories:

Comparison	Mitosis	Meiosis
Number of chromosomes in parent cell:		
Number of replications of chromosomes:		
Number of daughter cells produced:		
Number of chromosomes in daughter cells:		
Genetic relationship of parent to daughter cell:		

What's The Question?

Sexual reproduction involves the union of two different cells producing a new cell with unique DNA. Is this process the same in all animals? The answer, of course, is no - sexual reproduction is a complex process that is unique to each species. However, there are similarities throughout the animal kingdom. In this lab, look for those similarities and consider how the genetic information of parent cells is transferred to offspring cells.

1. *Conjugation in Paramecia*

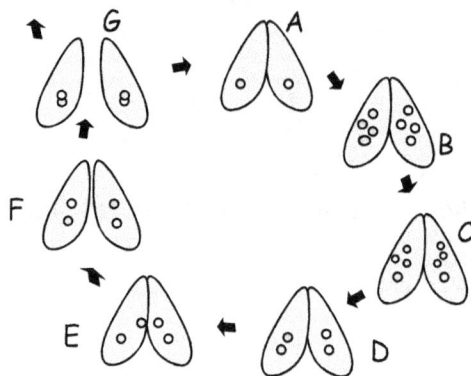

What is happening? Ask your teacher.

A

B

C

D

E

F

G

2. Reproduction in *Daphnia*

Daphnia magna (giant water flea) demonstrate forms of asexual and sexual reproduction. During summer, most are diploid (2N = 20) females that reproduce asexually. Each female carries up to 100 eggs that develop internally and eventually become adult diploid females. When faced with environmental stress, male *Daphnia* are produced from some eggs. These males fertilize specialized eggs that can withstand freezing and drying. These become next summer's *Daphnia* population.

Questions for later...

1. How many chromosomes are present in summer females?

2. How many chromosomes would be present in the gametes of ***Daphnia***? Explain.

Varieties of Reproduction

Name:
Date:

3. Reproduction in Earthworms
Earthworms can be described as hermaphrodites - that is, they contain both male and female parts. Fertilization is internal. During copulation, sperm is transferred from one worm to another, and both produce cocoons that contain fertilized eggs.

Questions

1. If earthworms adults have 8 chromosomes, how many would be present in their sperm and eggs?

2. Earthworms are not capable of self-fertilization? Why is this necessary to maintain genetic variability?

4. External Fertilization (Fish)
Many fish practice a form of sexual reproduction that involves external fertilization. In species like largemouth bass (*Micropterus salmoides*), the male prepares a nest and invites a female to lay her eggs in the nest. When she lays eggs, the male is stimulated to release sperm, resulting in fertilization. A female can lay as many as 110 000 eggs, but of these only 5-10 live to reach 25 cm in length.

Questions

1. If largemouth bass have 94 chromosomes in their body (somatic) cells, how many chromosomes are present in their sperm and eggs?

2. Do you think that all eggs produced by the female bass are fertilized by the male? Explain why or why not.

3. Why do you think bass lay so many eggs?

5. Internal Fertilization (Mammals)
Cottontail rabbits (*Sylvilagus floridanus*) are mammals that reproduce sexually by internal fertilization. These prolific breeders are known to produce 3 or 4 litters of 4-7 young per year.

Questions

1. How many eggs are produced by female rabbits each litter?

2. How many sperm are produced by male rabbits each litter?

3. Do you think that all eggs produced by the female rabbit are fertilized by the male? Explain why or why not.

4. Rabbits have 44 chromosomes in their somatic cells. How many chromosomes are present in the eggs and sperm of rabbits?

Explaining Reproduction

Act 4.4: Sexual Reproduction in Plants

Do you Remember? **List five examples of sexual reproduction in animals.** (Previous activity)

1. _____
2. _____
3. _____
4. _____
5. _____

What's The Question?

Algae and higher plants (i.e., those most changed or evolved) are considered to have evolved from algae that used flagellae to move about. In algae, there is very little difference in the appearance of male and female gametes. More advanced plants, such as mosses and ferns produce structurally different gametes, but retain the more primitive condition of alternation between vegetative and reproductive generations. In today's lab you should focus on similarities in the ways gametes are produced and become fertilized.

1. Conjugation in the Filamentous Alga (Spirogyra)

In this alga, whole cells act as isogametes (gametes that are structurally identical) in conjugation. Examine a prepared slide of conjugation in *Spirogyra*. The active (male?) cell moves across a conjugation bridge to fuse with the contents of the passive (female?) cell to produce a zygospore.

Diagram of Spirogyra (make your drawing here)	Questions
	1. *Spirogyra* is a filamentous alga that is said to be monoploid, that is, it has only one set of 24 chromosomes. How many chromosomes are present in each isogamete?
	2. How many chromosomes are present in the zygospore resulting from fusion of the two isogametes?
	3. Of what benefit is it to an individual *Spirogyra* to participate in this form of sexual reproduction?

Varieties of Reproduction

Name:

Date:

2. Life Cycle of a Moss

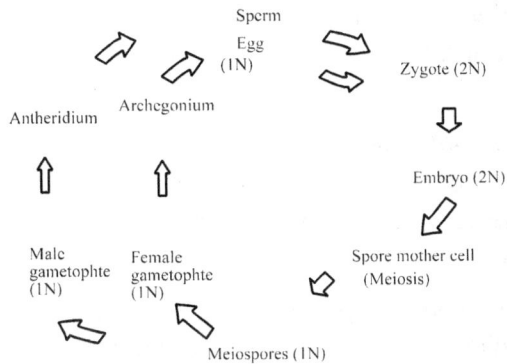

Life Cycle of a Moss

Sperm

Egg
(1N)

Zygote (2N)

Antheridium   Archegonium

Embryo (2N)

Male
gametophte
(1N)

Female
gametophte
(1N)

Spore mother cell
(Meiosis)

Meiospores (1N)

Why is this life cycle said to contain alternation of generations?

3. Life Cycle of a Fern

Compare this life cycle to that of mosses.

3. Life Cycle of a Gymnosperm

4. Life Cycle of an Angiosperm

Act. 4.5: The Human Female Fertility Cycle

Day in Cycle	1	2	3	4	5	6	7	8	9	10	11	12	13	14	15	16	17	18	19	20	21	22	23	24	25	26	27	28
F S H	50	61	72	78	83	77	79	84	71	60	50	98	72	50	50	45	44	46	42	39	36	33	30	32	34	37	36	42
L H	11	15	20	22	25	21	15	23	19	20	25	38	92	75	39	17	13	11	9	7	4	5	3	3	4	5	6	9
Estrogen	12	15	21	28	33	39	47	57	72	81	89	96	90	73	59	48	42	21	13	10	11	9	10	8	7	9	10	9
Progesterone	5	6	8	9	8	8	11	10	11	12	12	13	15	16	18	30	46	71	83	95	96	94	84	60	32	24	13	6

What's The Question? Every month, a woman's body undergoes a fertility cycle. This delicate, complex series of changes prepares her body to conceive and carry a healthy baby to term.
How does a woman's fertility cycle work?

What Are We Doing?
1.   There are four sex hormones circulating in a woman's bloodstream. Each hormone level rises and falls at certain times during her cycle. On the table at left are daily measurements of one woman's hormone levels.

2.   Use a pencil to plot the values of each hormone on the graphs at the back. Be very careful to check that you are putting the data in the correct graph.

3.   Connect the points with a smooth curve.

What Are We Thinking About?
1.   **F**ollicle **S**timulating **H**ormone (FSH) causes one follicle to grow. As it grows, it produces Estrogen, and eventually release an ovum.

2.   **L**uteinizing **H**ormone (LH) changes the follicle into a white body which then produces Progesterone.

3.   Estrogen generally increases fertility and prepares the body for pregnancy.

4.   Progesterone maintains the body's state of readiness for pregnancy. When the body is no longer prepared to become pregnant, the lining of the uterus is shed in the menstrual period.

Questions For Later...
1.   Once the ovum is released, it only lives for a few hours. If Juliet is going to become pregnant, when must Romeo's sperm cells be present?

2.   Juliet is much more than a collection of hormones. Her whole person is involved in this remarkable cycle. List some ways that Romeo could demonstrate respect for Juliet's whole person.

Varieties of Reproduction

Name:

Date:

The pituitary gland secretes two hormones. FSH stimulates the follicle, while LH makes the follicle produce progesterone.

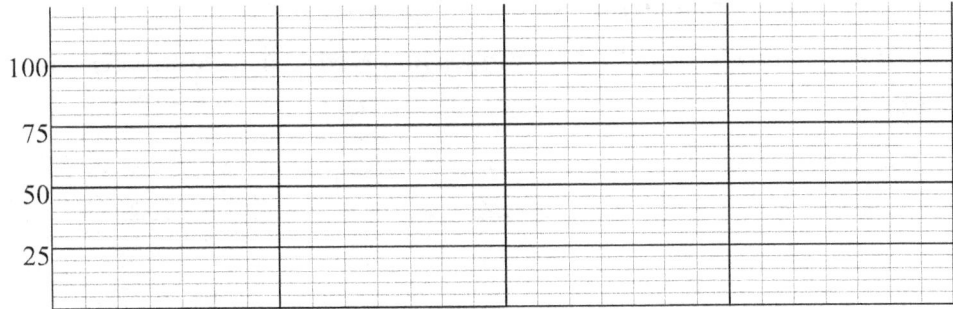

The Follicle undergoes changes, releasing an ovum and other hormones.

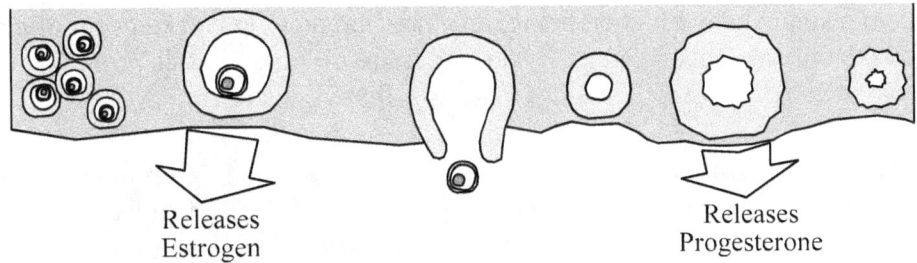

Releases
Estrogen

Releases
Progesterone

Estrogen and *Progesterone* are released by the follicle in the ovary.

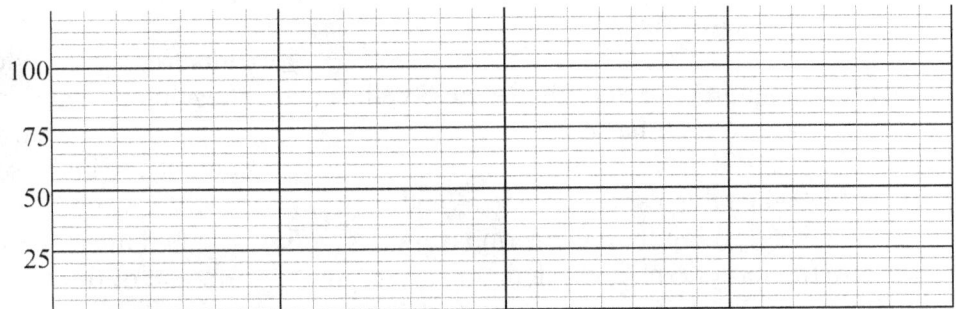

The lining of the uterus changes in response to the estrogen and progesterone.

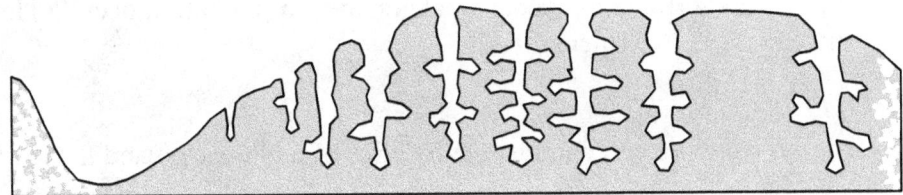

Some women experience emotional and other changes during the cycle.

Menstrual

Period

Feeling
of
well-being

Increased
need for
closeness

Acne

Anxiety

Irritability

Fluid retention

Wakefulness

Mood swings

Lab 4.6: Human Reproduction and Development

Do you Remember? **List the 4 hormones involved in a woman's fertility cycle and their function.**

1. _____

2. _____

3. _____

What's The Question?

From a single cell, a human being develops into a complex organism of perhaps 6 million million cells. How does this happen? What stages are involved in this process? What happens at birth? In today's activity, we will attempt to answer these questions.

What Are We Doing?

1. Obtain a prepared slide of cleavage in a zygote. Using a microscope or micro viewer, observe the size and general shape of the zygote during this stage.

2. Make a diagram of each of the 2-cell, 8-cell, blastocyst, and gastrula stages.

3. Complete the table listing the organs and tissues that develop from the germ layers: endoderm, mesoderm, and ectoderm.

4. Complete the table listing the main features of development that occur in each trimester of pregnancy in the space provided.

5. Identify 5 risk factors faced by the developing fetus and approximately when these risks are most critical during pregnancy in the space provided.

6. Explain the major events that occur and the amount of time involved in the three

Questions For Later...

1. Where does fertilization occur? Where does implantation occur? How much time passes between these two stages?

2. Briefly, compare and contrast the features of a blastocyst and those of a gastrula.

3. Where does the placenta develop and what is its function?

4. Why do babies that are born prematurely often experience difficulty in breathing? Use your understanding of development during the stages of pregnancy to help you answer this question.

5. Kangaroos are born after only 38 days. How is this possible? What differences exist between marsupials (*e.g.,* kangaroos) and placental mammals, like us?

Varieties of Reproduction

Name:

Date:

Zygote (2 cells)	Zygote (8 cells)	Blastula	Gastrula

Organ and Tissue Development in theThree Germ Layers	**Features of Development During Pregnancy**
Endoderm	First Trimester
Mesoderm	Second Trimester
Ectoderm	Third Trimester

Risk Factors Faced by the Developing Fetus	**Major Events During Three Stages of Birth Process**
1.	Stage Time (h) What is Happening?
2.	1st Dilation
3.	2nd Expulsion
4.	3rd Placental

Explaining Reproduction

The Genetic Code: The DNA Codons for Amino Acids.

In this modified genetic code, amino acids are coded by three nucleotides (a codon) in a sequence of DNA. Whenever these three nucleotides appear in this sequence, they form a kind of pattern which the indicated amino acid will "fit." In this way, DNA is a "pattern" for all of the molecules involved in life.

For example, suppose the base sequence AAA appears on your DNA. Your cells will always read the AAA as a call to add the amino acid lysine to the protein being built. The table below indicates which amino acid is called for by a particular three letter "code" on your DNA.

AAA K lysine	CAA Q glutamine	GAA E glutamic ac	TAA ● STOP
AAC N asparagine	CAC H histidine	GAC D aspartic ac	TAC Y tyrosine
AAG K lysine	CAG Q glutamine	GAG E glutamic ac	TAG ● STOP
AAT N asparagine	CAT H histidine	GAT D aspartic ac	TAT Y tyrosine
ACA T threonine	CCA P proline	GCA A alanine	TCA S serine
ACC T threonine	CCC P proline	GCC A alanine	TCC S serine
ACG T threonine	CCG P proline	GCG A alanine	TCG S serine
ACT T threonine	CCT P proline	GCT A alanine	TCT S serine
AGA R arginine	CGA G glycine	GGA R arginine	TGA ● STOP
AGC S serine	CGC G glycine	GGC R arginine	TGC C cysteine
AGG R arginine	CGG G glycine	GGG R arginine	TGG W tryptophan
AGT S serine	CGT G glycine	GGT R arginine	TGT C cysteine
ATA I isoleucine	CTA L leucine	GTA V valine	TTA L leucine
ATC I isoleucine	CTC L leucine	GTC V valine	TTC F phenylalanine
ATG M methionine	CTG L leucine	GTG V valine	TTG L leucine
ATT I isoleucine	CTT L leucine	GTT V valine	TTT F phenylalanine

Consider the sequence of bases below. What amino acid is coded for each group of three bases? Look them up on the table above.

DNA base	C	G	A	T	T	G	A	A	G	T	C	G
Amino acids												

Note that there are 64 possible combinations of the four bases, but there are only 20 different amino acids to code for. Therefore, some of the amino acids have more than one possible code in the DNA.

The proteins made up according to these patterns then go on to serve as catalysts for other reactions. For instance, one string of protein, perhaps200 amino acids long, may be able to bond to a sugar molecule and begin the process of breaking down the sugar to obtain energy. Another sequence of 300 amino acids may help you to fight off a virus. You can see why it is important that the DNA strand be copied accurately!

Do an hour every day...

The Five Day Project

Project 4.7: Reproductive Biology

Biology is undoubtedly the cutting edge science of twenty first century. This is the area in which you can quickly find new concepts being worked out, old theories being revised or discarded, new theories being developed, and radical work being done in every aspect.

The one common thread of this research is that it is founded upon the particle theory, and a theory of chemical change. How?

DNA is essentially a giant particle, a pattern that organizes the arrangement of other particles. The forces that hold the particles are electrical forces, and transformations that bind them together are chemical changes. Today, biology cannot be done outside these three understandings.

What's The Question?

We have spent an entire unit focussing on reproduction and the genetic material. In this project, you will demonstrate what you have learned by preparing a model of an application of reproductive technology. This project can take several forms. Make sure that you and your teacher are aware and agree upon the form your project will take.

What Are We Doing?   Here is a list of options:

1. Prepare a giant model of DNA to hang in your science room.

2. Prepare a detailed account of how genetic engineering technology is involved in some product or procedure, eg:
   Polymerase Chain Reaction (PCR)
   Genetically engineered herbicide resistance
   Genetic engineering to fight insect pests
   Use and production of genetically engineered Bovine Somatotropin (BST)
   The Human Genome Project
   Production of human vaccines
   Genetically "improved" crops

3. Prepare a detailed account of mutations that can eventually result in cancer.

4. The effects of one or more mutation causing (mutagenic) agents on cells.

5. The importance of Canadian research and technological development in genetics and reproductive biology.

6. Investigate careers that require an understanding of reproductive biology.

Do an hour every day...

The Five Day Project

Project 4.7: Reproductive Biology   Name:

0 You will produce a five page report. That's five pages of sentences, plus illustrations, cover and bibliography. Choose an Application of Genetics, preferably one being used near you. *Right now*, in one or two sentences, write down the topic question you have chosen to investigate. You must cover: *What*, using concepts from this unit *Where*, including facilities needed *Who*, including education needed *When* did this application begin *Societal implications* of this application.	**0** My Plan and Outline Date: / 4
1 *A list of sources* Prepare a list of at least 2 books and 2 internet sites to use for your project. Some examples are as follows. www.ericir.syr.edu/Projects www.biotech.chem.indianna.edu/pages/dictionary.html www.aba.asn.au/leaf2.html	**1** Prepare your bibliography now. You can always remove things that you did not find useful. Date: / 4
2 *Do the research.* Read and make notes summarizing your findings. Be sure to mark quotes exactly, and include the page number or page URL. Your notes ought to be at least 5 pages.	**2** My own notes, clear enough that the teacher can understand what I intend Date: / 4

3 *Create your rough draft* First you need a one page outline that organizes what you want to say, in the order you want to say it. Let a short sentence here stand for a full paragraph later. Second, write your paragraphs. You are trying to tell someone about your findings, so be sure to write clearly and define your words as you go along.	**3** My own notes, clear enough that the teacher can understand what I intend Date: / 4
4 *Proofread and edit* Work with a friend . Read each others' rough drafts, writing in questions and simple corrections with a red pen. Get your paper back, and re-write it to include the corrections.	**4** My own notes, clear enough that the teacher can understand what I intend Date: / 4
5 *Final project must include:* 1. 3 page body, 900 words 2. Diagrams, maps, charts, photos, graphs in addition to the body, in the order in which they are to be read. 3. Cover and Bibliography 4. This page, for teacher evaluation.	**5** My own notes, clear enough that the teacher can understand what I intend Date: / 4

Appendix: Laboratory Safety

The Hazards	*The Safe Way*
In this column is a list of lab safety issues that you will face in this course.	**Read this column to find out how to safely handle the laboratory problem.**
Eye Injury is possible from flying fragments of metal, glass or chemicals; from heat or flames; from caustic solutions such as acids or bases.	*Always wear safety glasses* in the laboratory. Never take your glasses off, even if you have finished your experiment. Other students may not have finished their lab work. The safety glass symbol indicates exercises in which safety glasses *must* be worn.
Crowding, Pushing and Horseplay increase the likelihood of a serious injury.	*Attend to your work.* Stay at the station you were assigned, so that there is room to work safely. If your teacher finds that your behaviour is a safety hazard, he or she may remove you from the lab. There is no place for behaviours which place others at risk of injury. Not at school, not at home and not at work.
Disorganized and Dirty Working Conditions are a hazard wherever they are found.	*Keep Lab Area Clean.* Clean and put away unused equipment. Tell your teacher about chipped, cracked, damaged or broken equipment. Do not leave anything on the floor, the desktop, the sink, or the cupboards that is not supposed to be there.
Broken Glass happens even to careful scientists.	*Do Not Touch* broken glass with your hands. Tell your teacher. When instructed to do so, use a broom to sweep the glass into a dustpan. Dispose of the broken glass in the special container provided. Do not leave it in the regular wastebasket: it could seriously injure a custodian.
Liquid Spills may consist of water, but they may also contain acids, bases, or toxic chemicals. You may not be able to tell the difference.	*Tell your teacher* about any spills immediately. Do not attempt to clean up without teacher instruction. Only if the teacher decides it's safe, use a cloth or paper towels to soak up excess liquid. Wipe the area clean with a damp cloth. Rinse the cloth frequently in fresh water. Wash your hands afterwards.
Solid Spills may consist of highly reactive chemicals. You may not know the specific hazards.	*Tell Your Teacher* about the spill, whether or not you caused it. Your teacher will instruct you on the safe way to handle the problem. In any case, the spill must be cleaned up promptly.

Appendix: Laboratory Safety

Name:

Date:

Open Flames are a frequent hazard. The Bunsen burner is the most likely safety hazard.	**Review Safe Handling of a Bunsen Burner** with your teacher. Be prepared to show how to light, operate and extinguish the burner at any time. Do not attempt to ignite pens, papers, rulers or other things. That kind of behaviour will certainly result in your being put out of the lab.
Fire. Any liquid solid or gaseous fuel burning where you do not want it to burn is a fire.	**Tell the teacher immediately!** Do not attempt to extinguish the fire with your hands, books, paper towels etc. Do not panic. Move away from the hazard. **Your teacher is the best judge of the appropriate course of action.**
Hot Metal or Glass cause more burns than any other hazard. There is usually no visible indication that they are hot. Glass in particular causes small, deep burns.	**Let Hot Objects Cool for 10 - 15 Minutes** before handling. Place all hot objects on a heat resistant pad. You and your partner will know where they are. Approach hot objects cautiously. Touch them at the coolest point first (the base of the retort rod, the bottom of the Bunsen burner or hot plate, the thumb screw of the iron ring). Use dry, not damp, paper towels to handle hot objects.
Hot Liquids such as boiling water or hot oil spread and splash rapidly. They also cling to skin and clothes.	**Let Hot Liquids Cool for 10 - 15 Minutes** before handling. Do not heat liquids in closed containers. Use hot plates rather than shaky retort rod assemblies. Do not heat more liquid than you need.
Obstructed Passageways prevent you from moving out the way of a spill or a fire.	**Stand at Your Lab Station.** Do not bring chairs or stools over to sit down. Your chair will prevent others from moving away from a spill or a fire.
Long Hair or Loose Clothing is more likely to become involved in your equipment. It can cause spills and breakage, or catch fire.	**Tie Back Long Hair; Secure Loose Clothing.** Outerwear in particular must be avoided in the lab situation. Jackets, sweat suits, hoods, etc are too large and awkward for the lab situation. They are also frequently made of materials that are flammable and can melt and stick to the skin in a fire. Avoid using laquer based hair sprays. A curly head of hair with hair spray can burn up completely in seconds.
Unauthorized Experiments can have unintended results.	**Stick to the plan.** Read instructions very carefully the night before the lab. Ask questions. Do not try experiments "just to see what happens." The dangers are too great.